M000249786

LIFE IN A ROMAN LEGIONARY FORTRESS

TIM COPELAND

AMBERLEY

Acknowledgements

For Genie and her Gamma, Anne

I have to thank Jeremy Knight and the late George C. Boon, both of whom took time to talk to the young boy who haunted their excavations. David Zienkiewicz provided much support in the writing of my *Isca Education Guide* (1993), the forerunner of this book. Mark Lewis, the curator of the National Roman Museum at Caerleon, discussed his ideas about aspects of the fortress of Isca and provided continual encouragement. Gill Dunn of Cheshire West and Chester Council provided me with access to images of the Chester fortress. Peter Guest provided sources and answered my many enquiries. Caro McIntosh kindly, as always, drew the maps and plans of the fortress. Louise Clough and Claire Shadwell helped me with updating my ancient historical skills on my laptop, and Dave Brookes produced miracles with my illustrations on his PC. Don Henson took photographs, and, unless designated otherwise, the colour images are all from the author's personal collection taken with the permission of CADW. The reconstruction of the Prysg Field barracks is courtesy of Julia Sorrell. Of course I take responsibility for all errors (and hopefully the joys) contained in this book.

First published 2014

Amberley Publishing
The Hill, Stroud, Gloucestershire, GL5 4EP
www.amberley-books.com

Copyright © Tim Copeland, 2014

The right of Tim Copeland to be identified as the Author of this work has been asserted in accordance with the Copyrights, Designs and Patents Act 1988.

ISBN 978 1 4456 4358 8 (print)
ISBN 978 1 4456 4393 9 (ebook)

British Library Cataloguing in Publication Data.
A catalogue record for this book is available from the British Library.

Typesetting by Amberley Publishing.
Printed in Great Britain.

Introduction

This book is largely based on the Roman fortress of Isca, the administrative headquarters of the Second Augustan Legion (*Legio II Augusta*) under the village of Caerleon – 'The Camp of the Legions' – near Newport in South Wales. I first visited Isca when I was eight years old. At ten, I was peeping over a fence at Jeremy Knight's excavation at Cold Bath Road and was invited to join his team as the pot washer, or as I like to think of it now, 'finds assistant'. One of my most formative experiences was Jeremy showing me around the preserved Prysg Field Roman legionary barracks in the south-west corner of the fortress, which remain the only examples on display in northern Europe, and seeing Alan Sorrell's reconstruction of them. Here I began my journey of understanding the archaeology of the everyday lives of legionary soldiers in their fortresses. It is from these barracks that the named individuals in this book originate, and it is also the part of the fortress from which they will lead their daily lives. Vital evidence will be used from the sites of the other permanent legionary fortresses in the Province of Britannia at Deva/Chester, the base of *Legio II Adiutrix* and later *Legio XX Valeria Victrix*, and at Eboracum, now present-day York, the home of *Legio IX Hispana* followed by *Legio VI Victrix*.

THE EVIDENCE BASE

Surprisingly, there is no surviving contemporary history of the Roman Imperial Army by an ancient author, and little contemporary, detailed examination of military practices. We have the evidence of only five writers, some of which might be applicable to the period of the Principate from 27 BC to AD 284, which is the focus of this book. The Greek historian Polybius in his *Histories* described the Roman army in the second century BC, but unfortunately this was before Augustus' reforms, which reshaped much of the military, including the reduction of the century to eighty men. Then there is Titus Flavius Josephus' *De Bello Judaico* (The Jewish War) *c*. AD 75, which extolled the achievements of the Roman army putting down a Jewish revolt in AD 66–70. In the late first or third century AD (it is contested), we have *De Munitionibus Castrorum* (About the Foundation of Military Camps), written by an unknown author, usually referred to as Hyginus or Pseudo-Hyginus. It is mainly about the building of campaign forts; however, some parts might be applicable to a permanent fortress. A Roman jurist, Publius Tarruntenus Paternus, wrote a four-volume work, *De Re Military* (On Military Matters), of which only two fragments survive in the *Digest of Justianian*, a compendium of Roman law compiled for the Roman emperor of that name (AD 530–533), and identifies the *immunes*, those who were excused from general tasks because of their specialist skills. Finally, there is Vegetius, a fourth century AD commentator who attempted to restore the Roman army to the military practices and virtues of the earlier years of the Empire with his *Epitomarei Militaris*, also referred to as the *De Re Militari* (The Military Institutions of the Romans), which is the most valuable source about the everyday military experience of the legionary. He probably also used Tarruntenus Paternus as a source.

Why is there so little? The reason is simple: the educated did not want to hear about the peacetime activities of the ordinary soldier, as they saw the army as a threat to their property – as it had been on many occasions – and wished the legions to be kept on the frontier, as far away as possible. They often saw the military as bridging the gap between barbarians and the Empire, and therefore of low status.

So, in literary terms, we are thrown back on the very scarce epigraphic evidence – inscriptions on monuments, dedications on altars, repetitive and formulaic lives on tombstones, graffiti on pottery and writing on wooden tablets. None of this evidence is dramatic, but it can get us closer to individuals. From the eastern part of the Empire, however, much was written about everyday life and duties of the Roman legionary on papyrus, which, due to the hot, dry climate, has survived. Vegetius tells us that the administration of the whole legion, whether services, military duties or financial transactions, was written down every day, including their daily peacetime guard duties and watches, and each man's leaves of absence and duration recorded. We are fortunate in having the identity of a centurion (we don't know the first part of his first name, his *praenomen*), Petronius Fortunatus, who served with the *Legio II Augusta* and later in Alexandria with *Legio III Cyrenaica*, and we can use this to demonstrate standard military organisation and practices across the Empire. While we know the names of a large number of centurions and their career paths, we know little of their everyday duties. Conversely, we can identify few legionaries, but know more of their everyday work.

In Britain, we are nearly wholly reliant on the archaeological evidence gained from excavation, field walking and surveying (especially with the recent increase in the use of geophysics at Caerleon) as our principle source of data to examine the predominantly anonymous people who lived in the barracks, and in the higher status areas of the legionary base. There are sound archaeological reasons for using Caerleon with its large, relatively recent and well-excavated sites, and the lack of buildings on an extensive area of Isca. However, Caerleon does have a rival in Inchtuthil in Scotland (its Latin name is not known), the northernmost fortress in the Empire. Here is a completely open site where the entire plan of the timber fortress has been revealed by aerial photography and trenching, which tells us a lot about how a fortress was laid out. However, the very short occupation, which probably only lasted for three years between

AD 83–86, and the limited excavation, has resulted in a lack of finds, which does not give us the detail for the current study. While these other bases will be used to give us a deeper understanding of daily life, it is Isca that will provide the unrivalled collection of objects that we need. The evidence from the other British fortresses suffers from being sealed below important medieval buildings, themselves of great historical value, and excavations and observations have been the result of opportunistic interventions before modern buildings or water pipes and electric cable have been laid. The evidence retrieved is still vital to our understanding of the working of fortresses, especially as these bases were laid out differently. There will also be excursions around the Empire, especially at Novae, the base of *Legio VIII Augusta* and *Legio I Italica*, in present-day Bulgaria, and *Legio III Augusta's* base at Lambaesis in North Africa.

Unfortunately, although we know the names of some of the soldiers who lived in the fortresses of Britannia, we do not have a commander's autobiography, the personal reminiscences of a centurion or a field recording of an ordinary legionary living in an alien landscape. George Boon, one of the most sensitive of the excavators at Isca, and who always encouraged the young who were serious about archaeology, including myself, said 'each generation builds its own fortress and each builds it differently'. So it is fitting I am 'building' this book on life in a Roman fortress in my own style, at the beginning of the twenty-first century.

A NOTE ON LATIN TERMS

It is difficult to know what an audience wants in terms of the use of Latin when describing aspects of Roman life in a military establishment. In the case of fortresses, some designations such as *rententura*, for the back area of the fortress, might have been used by those high up the command chain or architects who laid out the fortress, but rarely by the legionary, who was more concerned about his *contubernium* for eight men, or his barrack block, *hemistrigia*, the *thermae* for bathing and relaxation, or the *valetudinarium* when one of his mess mates was ill. The *canabae*, the civilian settlement outside the fortress walls, was also important when he needed more food, to worship his favourite god, or simply to find a woman. However, if the reader wants to appreciate the actions of individuals and their experience of the fortress, then it will be necessary to speak some of the language, even if we don't know 'camp' Latin (*sermo militaris*), which was in general use.

I

The Roman Army

During the Roman Republican period, temporary armies were raised to deal with a specific hostility, and a high proportion of the troops were expected to be released at the end of the emergency. It was only at the beginning of Augustus' Emperorship in 27 BC that a permanent, voluntary army was developed; this was further refined during the period commonly known as the 'Principate', which lasted until AD 284. The number of troops in this stable, imperial Roman army may have been small compared to that of the population of the Empire, but it was seen as more valuable to have a limited number of high quality, well trained, content troops, than the possibility of large numbers of unfit, begrudging, raw recruits and veterans. The permanency of this army also produced an enforced hierarchy of ranks with set terms of pay and service, acting under rules of military disciple with standardised equipment, training and shared battle tactics. This standing army comprised two categories of troops, tasked with the annexation and security of a growing empire. Understanding the subsequent roles of these fighting men is essential to gain an insight into the variety of their bases.

Legionaries (*milites*) and individual soldiers (*miles*) were the heavy infantry largely based in fortresses. They were Roman citizens and served for twenty-five years, before getting a grant of land or of money *in lieu*. Auxiliary troops were the standing non-citizen corps of the army, and being free provincial subjects of the Roman Empire, they usually maintained their identities of place and origin in the names of their regiments. Auxiliaries served for twenty-five years, after which they were granted Roman citizenship. The *Ala I Thracum*, from Thrace in south-east Europe, had a close relationship with *Legio II Augusta*. However, the relationship between the legionaries and auxiliaries is often not a straightforward one, especially when on campaign.

VEXILLATION FORTRESSES

During a campaign, the heavy infantry of the legion could act with lightly armed or mounted auxiliary regiments to provide the best possible combination of troops for a specific landscape and the type and strength of the opposing forces. Fighting against opposition that used guerrilla tactics required a strategy that would allow for control of the valleys and high ground around them, and probably involved swift, mounted auxiliary horsemen, whereas if the enemy felt confident to come into the open, a more massive force, such as a legion or part of one, would be needed to deal with the threat. It is highly likely that these vexillation formations were the rule and not the exception during the early, restless period of the conquest, parts of the 'legion' with its auxiliaries being in several locations, each forming a 'battle group'. These military groups were housed in, or around, 'vexillation fortresses', which are best seen as the places for the assembly of operational battle groups, and may also have been used as campaign winter quarters. There are fourteen known or suspected vexillation fortresses in the province of Britannia, and these provide evidence of the progress of the conquest after the landing of AD 43.

LEGIONARY FORTRESSES

We know of sixty-five fortresses in the Roman Empire, not all occupied at the same time by the thirty legions, which dominated the military landscape of the Roman Empire frontiers, each one the administrative base for (usually) a single legion. These legions are often described as the backbone of the army, and were formed of around 5,500 military personnel (we are not sure of the exact figure as it varied unit to unit). They were arranged in nine cohorts of 480 men, allocated to six centuries of eighty men – except for the first cohort, which comprised of double centuries of 160 men. Of the known fortresses, not all were permanently occupied; some were temporary operational bases abandoned as the conquest moved onward. These early fortresses tend to be focused in areas where the indigenous peoples had not been completely subdued; in Britannia, the fortresses at Lincoln, Exeter and Gloucester were of this type. These bases were never intended to be permanent installations (the building of a labour- and resource-intensive stone wall was not undertaken), and as it was not known when they would become strategically redundant. Timber was strapped to the inner clay of a rampart, therefore ensuring stability for as long a time as possible. Finally, when it was clear that a province was safely part of the Empire, stone fortresses were constructed as permanent bases at Isca/Caerleon, Deva/Chester, and Eboracum/York. They acted only as administrative centres while parts of the legion were on duties elsewhere in, or outside, the province.

FORTS

Forts were occupied by the auxiliary units of a legion – if the legions were the backbone of the Roman army, then these troops were certainly its limbs. Auxiliary troops were organised in units known as *cohortes* (infantry) and *alae* (cavalry), or a combination of both (*cohortes equitatae*), each usually 500 men strong. The forts were situated at key positions along a network of roads built by the legions, and which effectively smothered the area, enabling rapid mutual support from other units in the case of insurrections. In the early days of the annexation of territory, or later rebellions, the forts were part of an internal security strategy. In more peaceful times, the roles undertaken were more appropriate to the control of local populations, including administrative tasks such as tax collection. As the Roman army successfully stifled any opposition, the number of forts decreased and their buildings were demolished, or the area of the fort diminished. However, it is not unusual to find that some locations were the sites of a succession of forts built, demolished and reoccupied as needed, or the area of an existing fort increased, sometimes with considerable time between these events. Frustratingly, we have no idea of the wider political or military reasons that resulted in these actions.

Permanent fortresses and forts were usually constructed after the campaign's successful conclusion, as it was only after enemy defeat that men could be made available to build these stations and infrastructure. The forts might be seen as an extension of the fortress; the legate of the legion being responsible for the whole command. The relationship with a particular legion was an enduring one, possibly the result of the formation of the battle groups during the invasion period. We actually have literary evidence that it was the legionaries who built the forts, or at least the defences, and they were protected by the auxiliary troops. This demonstrates the mutually beneficial speed of response from lightly armed auxiliary troops or cavalry, and the specific engineering skills of the heavy infantry (Hanson, 2009). During the early days of the occupation, the bulk of the auxiliaries were stationed in upland areas with subsistence economies incapable of providing sufficient foodstuffs and supplies. This was especially so in areas where the population had been largely destroyed by military action. Consequently, food and other supplies were transported from the parent fortress, until some measure of self-sufficiency was established, thereby making the fort independent of the demands of the agricultural year. Much of the hardware needed by an active fort – military equipment, leather goods, and tools – would have been supplied by the *fabricae* (workshops) of the legionary base. There were probably reciprocal arrangements, such as the auxiliary forts who supplied cured hides from their own herds of cattle for tanning and the manufacture of a wide range of military items.

THE CONQUEST OF BRITANNIA

Clearly, a legionary's life is going to be different at each stage of a provinces' conquest – campaigning, annexation and permanent presence – so it is valuable to use the role of *Legio II Augusta* as an exemplar for these phases, until its arrival at Isca.

The Britons had never faced a sophisticated army like that of the Romans, and they were reliant on groups of warriors raised at times of crisis, preferably outside of the agricultural season. Goscinny and Uderzo (1966) in *Asterix apud Britannos* (Asterix in Britain) suggested that the Britons stopped fighting at tea time and weekends, and that the Romans fought only at tea time and weekends – but that is very much a French view! The Britons fought with determination and resilience, but faced a superior, professional army, and it is likely that between 100,000 and 250,000 may have perished in the conquest period. Eventually, the British military garrison was the largest of all the provinces in the Empire, and one of the highest densities of military personnel – 10–12 per cent of the army in 4 per cent of its territory – and in the second century AD there were about 55,000 Roman troops in Britannia.

There is a debate about the location of the landings of AD 43 – were they at Richborough in Kent, or in the Solent on the South Coast, or both? The literary sources are ambiguous, and archaeology is unable to pinpoint events within such a narrow timescale. However, it would seem that *Legio II Augusta*, which had been moved from Strasbourg for the invasion, was in operation south of the Thames under the command of Vespasian (the later Emperor), and it is an attractive proposition to presume that the legion did disembark from the Solent. We know from the Roman historian Suetonius that Vespasian, with *Legio II Augusta*, conquered the Isle of Wight with the help of the navy, and took a large number of hill forts in the Dorset area. The archaeology suggests that there were troops near Chichester, which has been dated *c.* AD 45–48, and from *c.* AD 48–60 there also appears to have been a vexillation fortress at Lake Farm, near Wimborne, Dorset. It has been proposed that in the mid-forties until the mid-fifties, part of a vexillation of the legion might also have been at Alchester, Oxfordshire. A fortress in Exeter, founded *c.* AD 60, was an earlier Isca, on a narrow spur of land above the River Exe (identified as Isca/Exeter here, as Isca will always refer to the

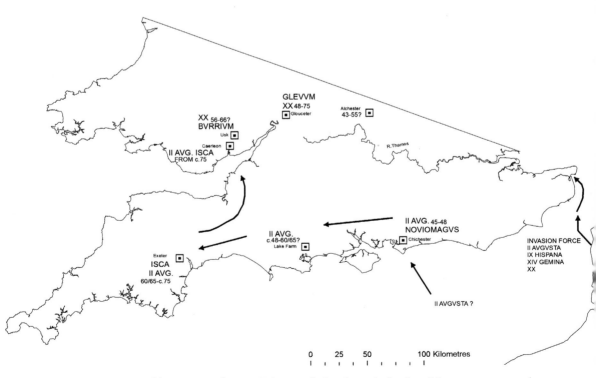

The map gives a possible progress of *Legio II Augusta* before its arrival at Isca. However, we can only be certain of its presence at Exeter, although even there the dating is difficult. (*Caro McIntosh*)

Caerleon base). It was certainly built by *Legio II Augusta* as tiles produced by the same mould have been recovered at both sites. Although the fortress was relatively small, some 16.4 hectares, it appears to have had all the components expected in a legionary base, but in a more limited area. It can be assumed that vexillations of legionaries and auxiliaries would have been operating throughout the South West as far as Cornwall, building the Fosse Way – a major route running to the North East across the nascent province – and planting forts along the route to control newly annexed areas. A vexillation from the legion was present in AD 69 at Cremona in northern Italy at the Battle of Bedriacum, when the 'Year of the Four Emperors' ended with Vespasian's victory and the vanquishing of his rivals. These fortresses were constructed to control specific peoples, such as the Silures from the bases of Glevum/Gloucester, and the Dumnonii and Durotriges from Isca/Exeter, and indeed several became the administrative centres, capitals of these large communities. However, it is possible that the permanent fortresses of Isca, Deva, Eboracum, and the short-lived, anonymous base at Inchtuthil in Scotland, were the culmination of a long-term plan conceived by the Emperor Vespasian (AD 69–79), using the experience and knowledge he had gained as commander of *Legio II Augusta* during the invasion period.

At the end of the first century AD, there were around 10,000 troops in the Isca command – 5,500 legionaries and a similar number being auxiliaries. If we add a comparable number for the Deva command, it would appear that there would have been between 25,000 and 30,000 soldiers for a population between 330,000 and 500,000 in the area of present-day Wales, which would represent a ratio between 1:11 and 1:17 of troops to civilians. It is highly likely that *Legio II Augusta* was also charged to supervise South West England, which it had very recently left. With these areas of the province secure, the fortress could act as an administrative base, and a reservoir of men who could be sent elsewhere when needed.

The role of the legion at this time is summed up in a speech to *Legio III Augusta* at Lambaesis, in northern Africa, when the Emperor Hadrian (AD 76–138) commented on the move of the unit south from its previous base at Ammaedara:

> Catullinus, my legate, is keen in your support; indeed, everything that you might have had to put to me he has himself told me on your behalf: that a cohort is away because, taking turns, one is sent every year to the staff of the Proconsul; that two years ago you gave a cohort and five men from each *centuria* to the fellow third legion, that many and far-flung outposts keep you scattered, that twice within our memory you have not only changed fortresses but built new ones.
>
> (*Translation by Speidel, 2006*)

Just twenty years after constructing Isca/Exeter, *c.* AD 75, *Legio II Augusta* had to build a new fortress on the banks of the River Usk ('Isca' probably being the British-Celtic name for water).

THE ESTABLISHMENT OF A ROMAN LEGIONARY FORTRESS

If the broader political and strategic demands determined the area where a fortress was to be placed, its physical location in the landscape was the result of tactical and practical decisions. There needed to be sufficient local resources – especially stone and timber – and, if possible, a river that was navigable and could provide water until the building programme included an aqueduct. The site needed to have a low enough water table to avoid flooding, yet high enough to provide water from wells if necessary. When fully functional, the fortress would have used huge amounts of water daily, an estimated 2.37 million litres at Deva, and so it needed to be on slightly sloping ground to facilitate drainage by gravity into the river. A legionary base needed considerably level land around it for the construction of camps, parade grounds and eventually the civilian settlements that were under the authority of the legionary legate. Although Isca was established on a virgin site (an earlier fortification has not yet been found), the location of the base at Deva appears to have had two military installations before the construction of the fortress, demonstrating the fitness for purpose of the location.

Hyginus recommended that camps should not be overlooked by hills, as the enemy might be able to view the base and have the advantage of attacking downhill. The Isca/Exeter fortress was on a rocky spur above the Exe, as was Deva above the Dee. However, Isca is overlooked on three sides, and even though sited on a river terrace, it is in a position that could not be easily defended. This indicates that the communication advantage of being in the river valley outweighed the defensive factors, although there was unlikely to be a local threat when the fortress was founded, even if there was still active campaigning in its hinterland. The centre of the Welsh mountains massif, the steep valleys of the rivers that drained them, and the flat platform of the coast through which these water courses poured into the Bristol Channel, were the sites of auxiliary forts that could be easily reached from Isca. The River Usk provided a water route to the South West via the Severn Estuary and, as important, fast and well-made roads connected the fortress to the Midlands, northern England and London, which gave opportunities of high mobility for the legion if necessary. The detail of building a military base on campaign is described by Vegetius, and presumably reflects some of the aspects of the physical construction of a permanent legionary base.

BUILDING THE DEFENCES

> After the ground is marked out by the proper officers, each century receives a certain number of feet to entrench.
>
> Vegetius

One of the first actions at any fortress site was to build the defences to secure it – to entrench the area – and this implies that these proper officers, the *agrimensores* (the measurers of the land), carried out the initial surveying of the fortress and set the position of the walls and gateways. Hyginus states that a camp should be 3x2 in

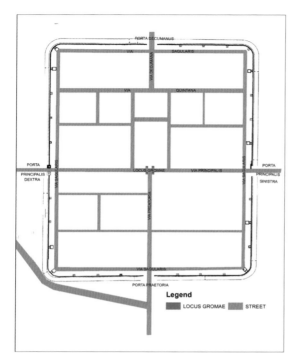

The basic layout of the fortress includes gates and roads and the *locus gromae* from which their position was established. Missing from the plan are the names of the three zones: the *praetentura*, between the *via principia* and the *porta praetoria*; the *latera praetoria* was the central zone between the *via principalis* and the *via quintana*; and the remaining zone at the rear of the fortress was the *rententura*. (*Caro McIntosh*)

proportion, so that a blowing breeze can refresh the army. If it was longer, the trumpet call could be sounded, but in a disturbance the horn could not be easily heard at the *porta decumana*. Isca is 490 metres from north-west to south-east, and 418 metres from south-west to north-east, covering 20.5 hectares in a 7x6 ratio and forming a 'playing card' configuration. Hyginus also tells us that the starting point for construction was at the entrance to the *praetorium*, at a position named the *locus gromae*. This point was where the *groma*, the Roman iron-footed surveying instrument, was put, so that by sighting from this point the gates of the camp would form a star. In a marching camp, the *praetorium* was the legate's accommodation, as well as being where decisions were made. In a legionary fortress, it became the personal accommodation of the commander, and the *principia* was the headquarters of the legion. Once the sites of the gates had been established, the fortifications could be constructed to link them. These defences comprised a patrol road on the edge of a ditch – the ditch, the turf rampart, a length of flat ground (*intervallum*), and the street running around and below the walls (*via sagularis*). This depth of protection would ensure that no projectile from an enemy could reach the barrack blocks lying along the street.

In a campaign fortress, setting out the defences was urgent and would haven take all the available manpower for a short burst of effort. Even in an area already under Roman control, which surely must have been the case at Isca, with the Silurian people cowed and few men of fighting age surviving the subjugation, protecting the legion would still have been a priority. The ground had the turf removed, and a rampart of clay was constructed on a 'timber raft' of brushwood and branches with lenses of turf

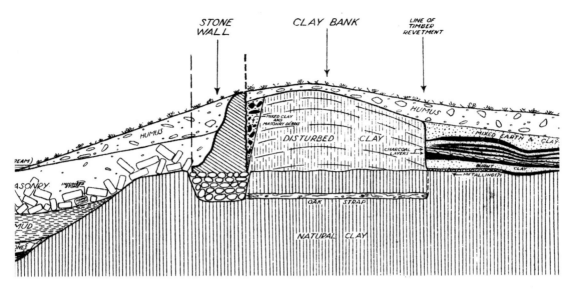

This section from the Prysg Field excavations shows how the clay rampart was cut back to receive the stone wall. The position of the timber breastwork was probably above the middle of the stone wall; however, the front of the bank would have sloped slightly to give stability. (*Archaeol Cambrensis 1931*, Vol. 86, Fig. 3)

embedded to provide stability. This was enhanced by a turf-faced wall added to the front, and a revetment located at the rear to stop the clay, dug out of the ditch, from slumping in wet conditions. A stout fence of timber posts was constructed along the top of the rampart, just behind the front of the turf wall.

There is little convincing evidence of timber towers along the rampart at Isca, although they may have been destroyed by subsequent stone built ones on the same footprint.

The arrested development of the Inchtuthil fortress is invaluable in illustrating the early growth of the other permanent bases. It is also of particular value to our understanding of Isca, as the length of the perimeter wall at Inchtuthil was 1,840 metres and at Isca it was 1,852 metres. Elizabeth Shirley, an archaeologist and quantity surveyor, attempted to quantify the labour, materials and time taken in the construction of the timber defences at Inchtuthil. She estimated that digging the ditch and constructing the bank would have taken about 512,000 man hours – 1,000 men working sixty-four days – this would mean that the equivalent of only thirteen centuries, a fifth of the legion, might be involved. The construction of the rampart would have been undertaken by the *milites* (the common foot soldiers) and slaves or captives taken on campaign being supervised by *immunes* (legionaries expert in construction, such as specialist carpenters). The stone wall was probably inserted in the second season, and being fire-resistant, hard-wearing and having a longer life than timber, symbolised permanency. The importance of making such a statement was not just a matter of demonstrating power and prestige to the local people, but also of reminding the troops

The stone-built fortress wall at York. The tile course is to ensure that there is a level base for the stone above, and, if the wall was seen and not covered with plaster, a decorative device. (*Don Henson*)

An interval turret at York. It is not bonded into the fortress wall and is likely to have been added much later. The arches above are supporting the Victorian walkway that went around the medieval city wall. (*Don Henson*)

of their successes in getting to this location, in what was enemy territory, and finishing the task of conquest and annexation.

The addition of the stone circuit may have taken 400,000 man hours – 500 men working 100 days – depending on materials and weather conditions. Inchtuthil did not have any interval towers, either of timber or stone. There is good evidence from other military sites that the outside face of the wall was plastered in white, with false joints outlined in red, to give an impressively even look. Again this might not be just for the benefit of the local population, but also for the prestige of the troops. It is possible that the process of inserting a stone wall in front of the rampart and building the turrets was begun within a few seasons of the establishment of Isca. However, a coin from AD 87 was found below a stone turret at the south-east corner at Isca suggests that it is more likely to have been in the last decade of the first century AD. This is not to say that the whole fortress perimeter was constructed in stone at the same time, as at Deva and Eboracum, and it seems that some parts of their defences were not completed until decades after others.

When the stone defences were completed, there were thirty turrets along the wall, about 43 metres apart, in the fortress corners, as well as eight gate towers. The interval between these towers would ensure that every part of the rampart was within a javelin throw. The *via sagularis* ran below the defences, which allowed the troops to reach any point of the defences quickly.

> The first thing to be done after entrenching the camp is to plant the ensigns, held by the soldiers in the highest veneration and respect, in their proper places.
>
> Vegetius

Within the standard layout of the streets structuring every legionary fortress, there seems to have been no rigid plan or set of fixed dimensions, and the layouts of individual buildings were probably the result of decisions by the architects. These architects served the legion, but also responded to their own experience or the wishes of the camp prefect, who may not have been the same one at previous fortresses. What is clear is that there was no standard plan associated with an individual legion – *Legio II Augusta's* fortresses at Strasbourg, Exeter and Caerleon all have different layouts, responding to diverse landscapes and the stage of conquest or defence of a province.

The *agrimensores* carried out the initial surveying of the fortress and the subdivision into plots, with the detailed laying out of the main buildings being undertaken by the *mensores* using the Roman foot *pes monetalis* (pM) – 1 pM equalling 296 mm. The key structure was the headquarters building, the *principia*, which was placed at the centre of the fortress as far away from the defences as possible, as it held the sacred standards, especially the legion's eagle, which embodied its identity. The orientation of the *principia* had determined the location of the gates, and therefore the defences, and it would also establish the whole layout of the fortress. This was always determined from the rear, so the *via principalis* ran across the frontage of the *principia* and between *porta principalis dextra* (right side) and *porta principalis sinistra* (left side). The street was lined with the important high-status structures of rank and administration. Hyginus

The turret inside the south-eastern corner at Caerleon. Although a straight joint between the wall and the turret would seem to indicate the former was earlier, the stones of the defensive wall were larger than those of the tower and therefore it would have been hard to bond them together, suggesting that both are contemporary. The offset on the inside of the fortress wall indicates the level of the wall walk. (*Author's collection*)

tells us that the standards in the *principia* 'should look down' the *via praetoria*, which led to the main gate (*porta praetoria*). These main streets came together outside the *principia*, where there was a four-way triumphal arch placed on the site of the *locus gromae*, perhaps commemorating victory over the Silures, or even the whole region of South Wales. All the other roads in the fortress were laid out on a grid based on these two main thoroughfares. The rearmost portion of the fortress was divided into two by the *via decumana*, which continued the line of the *via praetoria* and ended at the fourth gateway, the *porta decumana*, with the *via quintana* cutting across it and dividing the rear from the central portions of the fortress. These roads divided the fortress into three zones: the *praetentura* between the *via principia* and the *porta praetoria*; the *latera praetoria* was the central zone between the *via principali* and the *via quintana*; and the remaining zone at the rear of the fortress was the *rententura*. Within these zones were the *insulae* (islands) where buildings for different ranks and functions were sited.

> After this the *praetorium* is prepared for the general and his lieutenants, and the tents pitched for the tribunes, who have soldiers particularly appointed for that service and to fetch their water, wood, and forage. Then the legions and auxiliaries, cavalry

and infantry, have the ground distributed to them to pitch their tents according to the rank of the several corps.

Vegetius

In a legionary fortress, the position of its accommodation was based on the criteria of rank and status, but there were also other factors resulting from the semi-permanency of the operational or the permanency of the stone-built administrative bases for a legion. There was a tight community founded on the all-encompassing routine of camp life and an identity focused on the Emperor and Roman state, which was seen in ritual practice intimately connected to symbols, such as the legion's god-eagle, which was kept in the building at the centre of the fortress. The army was a bureaucratic organisation with an insatiable appetite for form filling and reporting, which required a large number of spaces for administration, not only for matters pertaining to the fortress and its garrison, but also its role acting as the centre of a large estate belonging to the legion (*prata*). There needed to be a vibrant craft industry to supply the legion and its auxiliary troops with weapons, and this resulted in an array of workshops as well stores, especially for foodstuffs. Finally, and of huge importance, was keeping the troops well-housed, healthy and entertained, which would avoid mutiny. Permanency also dictated that the garrison of Roman citizens had to have its own social and cultural life isolated from the population that it controlled, which took the form of the baths, the ampitheatre, and the civil settlement outside the walls supervised by the Legate.

The *via principalis* had the *principia* at its middle point. In the *insula* beyond, on the same side, was the *praetorium*, a large 'palace' fitting accommodation for the *legatus Augusti pro praetore*. On the opposite side of the *via principalis* was the s*camnum tribunorum* (the houses of the six tribunes), all undertaking military service before embarking on a public career either as a senator or equestrian. The third in command was the *praefectus castrorum* (prefect of the camp), and was responsible for the administration of the fortress, its supplies and logistics, as well as building the auxiliary forts. In the *insula*, next to the *principia* and stretching to the *porta principalis dextra*, were the barracks of the First Cohort, five double centuries (800 men), whose centurions were of higher rank that the others in the fortress and known as *primi ordines*.

Around the defences were the other nine cohorts, each comprising of six centuries of 80 men. The position of the barracks was not just to access the rampart easily in case of trouble from outside (or inside) the fortress, but also to be as far away as possible from the ceremonial centre of the fortress. It has been suggested that the *praetentura* was the most prestigious position for senior cohorts, close to the *porta praetoria*, while the *rententura*, the location of the Prysg Field barracks, was for the most junior. This might be reflected in the type of role the cohorts played in the fortress life, as the junior ones would have been near, and downwind, of the *fabricae*.

If the accommodation for each of the ranks from legate to legionary is predictable, it is usually the location of service and communal structures that are unique to each fortress. While the bath building (*thermae*) was usually near the heart of the base, and the granaries near to the gate closest to the river on which the heavy commodity was usually conveyed, there was no common template. Using only archaeological evidence,

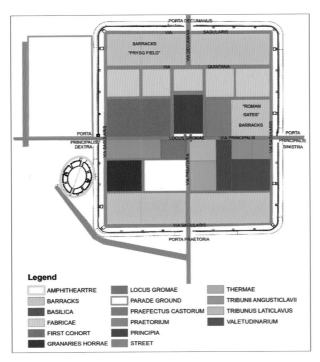

Legend

▢ AMPHITHEARTRE	▨ LOCUS GROMAE	▨ THERMAE
▨ BARRACKS	▢ PARADE GROUND	▨ TRIBUNII ANGUSTICLAVII
▨ BASILICA	▨ PRAEFECTUS CASTORUM	▨ TRIBUNUS LATICLAVUS
▨ FABRICAE	▨ PRAETORIUM	▨ VALETUDINARIUM
▨ FIRST COHORT	▨ PRINCIPIA	
▨ GRANARIES HORREA	▨ STREET	

Within the three zones of the fortress were the *insulae* (islands) where buildings for different ranks and functions were sited. The blank *insulae* has recently been shown to have been, at least in part, a store. (*Caro McIntosh*)

identifying hospitals, stores and workshops that filled the remaining space is fraught with problems.

THE LOGISTICS OF CONSTRUCTING THE FORTRESS

Constructing a legionary fortress was a considerable undertaking, requiring the supply of vast quantities of materials, labour, substantial resources and organisational abilities. The enormity of the task has been highlighted by Elizabeth Shirley's calculations with the timber fortress at Inchtuthil. Inside the defences were a small *principia*, large hospital, a workshop, four tribune's houses, six granaries, sixty-four barracks with centurions' quarters, and over 170 storerooms, together with roads, ovens and drains. The fortress lacked only a legionary bathhouse (*praetorium*), three or four tribunes' houses and two granaries. Site preparation had to be thorough and was back-breaking. The area had to be drained if wet and levelled using clay – the preferred building material of the army – and at the same time as the defences were being constructed, a drainage system was laid. Raw materials had to be extracted and processed, components manufactured, the legionaries trained, cared for and organised on a day-to-day basis. Tools, presumably, were transported with the legion, as was constructional equipment, though some will have been made onsite. It was suggested that the ground works – excavation and backfill of all wall trenches, post holes, the collection of timber, preparation and placing and fixing it into frames – might have taken 628, 000 man hours. Shirley estimated that 1,600 cubic metres of timber had been used, weighing 16,800 tonnes. For roof coverings of tile, some 366,000 *tegulae* (flat tiles) and 375,000 *imbrices* (half-round tiles placed over the cavity between *tegulae*) were manufactured, and some 776, 000 wooden shingles had to be produced, transported and fixed. Altogether, 1,555,000 nails had been manufactured, and during the archaeological excavation of the fortress, 1 million showing signs of extraction were recovered from a pit, probably buried to deny the metal for the enemy to use for weapons after the abandonment of the fortress. The buildings within the base would have taken somewhere in the region of 337,000 working days, or 5,000 men working for 2¼ months. These figures do not include the non-local preparation and manufacture of materials, as in the case of producing the tiles (extracting and processing the clay and its admixtures), the provision of water, tools and fuel, and the construction of the kilns, as well as transport. Many of the materials would have been more locally sourced, such as withies or wattle for walls, and green timber to be split for shingles. Elizabeth Shirley suggests that at Inchtuthil the construction work was probably spread over two seasons, with the first concerned with securing the site – building turf rampart and ditch, most of the barracks and centurions' quarters and some granaries. The second season completed the accommodation and other granaries were constructed, as well as the hospital, the headquarters, a workshop and *tabernae* (small rooms used for variety of purposes), and the stone defensive wall. Presumably, the main bathhouse and aqueduct were probably to be built in the third season.

While the construction of the fortress took place, other supporting tasks needed to be completed – staffing the supply compound, including the preparation of food, cooking, clothing, bringing water to the site, caring for and moving animals, moving equipment and constructing tools and plants, as well as the entire organisation, administration and overseeing of the work. Of course some legionaries would be on guard and protection duties while the fortress was in such a vulnerable state. Other legionaries might be engaged on essential military training and practice away from the base, especially as the auxiliary forts to protect the route to the fortress needed to be constructed.

The most accurate dating of the beginning of the construction at Isca is provided from a series of thirteen timbers used as piles over a well in a tribune's house, and had a single tree-felling date between AD 71/72 and 73/74. That these timbers may have been cut down in winter would not be surprising as oak, one of the most commonly utilised woods, is easier to use when green and splits if it dries out too quickly. Similarly, turf is much easier to cut and lift when it is not growing. So the construction season might have been in autumn and winter, after the campaign season in spring and summer. Although the accepted date of the foundation of the fortress is AD 75, there is no reason why advance parties could not have begun the work earlier.

LIFE IN THE FORTRESS WHEN?

It is unlikely that neither timber- nor stone-built fortresses were ever full of troops, and it is possible that a legionary serving twenty-five years may never have seen his home base, or indeed military action. The ebb and flow of a garrison over decades could be based on the gravitational pull of an emperor's aspirations. To satisfy those ambitions, major parts of the legion could serve elsewhere, undertaking civil engineering projects or even expeditions on other borders of the empire. So, trying to write about life in an established legionary base comes down to the question of 'when?' It is hugely difficult to relate archaeological evidence to historical events, but it is worth trying because it demonstrates the dynamic nature of a legion and its fortress and is a matter of finding an indicator that identifies periods of occupation and absence. The most reliable informal measurement tool would be a communal amenity such as the fortress baths or the amphitheatre, as both needed large numbers of men to build, repair and renew them, and in the case of the bath building, to provide the massive amounts of fuel needed to run it. If bathing was considered to be an everyday part of hygiene, then the smaller baths outside the fortress would have been used by the nominal group of men acting as caretakers when most of the legion was absent.

Let's examine the baths at Isca based on the chronology provided by David Zienkiewicz:

i)   They were finished AD 77/78, and work was begun on the *basilica thermarum*, but halted. Was this because the legion (or vexillations from it) was fighting in Agricola's Northern Wars between AD 78 and 80 and thus away from the fortress?

ii)   Repairs and alterations to the baths indicate that the full legion was back in base by AD 100. However, in *c.* AD 120–40, Hadrian's and the Antonine Walls were being built with the legion providing the part of the labour force.

iii) Around AD 150, the legion was back at Isca and during the AD 150s and 160s, the baths were repaired and altered (although during that period there was further trouble in the North and there were wars in Armenia in which British legions fought, and the German Danube frontier needed strengthening).

iv)   Sometime between AD 170 and 220, the baths were dismantled and then restored again, as though an order had been given to move the legion that was then rescinded for some reason, probably because circumstances in the wider empire changed. Around AD 200, the fortress was being rebuilt, but then in AD 208–11 there was Severus' northern campaign, of which the legion was a part. By AD 230, the baths were out of use and mothballed, although Isca remained a notional base and was abandoned *c.* AD 300. There would appear to be just two periods when we can be reasonably sure that the legion, or significant part of it, was housed in the fortress: AD 100–200.

Although modern pictorial reconstructions of fortresses show the full complement of functioning barracks around the walls, in reality there would have been some gaps. In AD 100, a *miles* on guard duty on the rampart above the Prysg Field would have seen a mixture of barracks – some converted to stone and others still in timber. Barrack XII of the Prysg Field sequence excavated at Alstone Cottage was not yet built, and the space was occupied by a structure with a concrete floor and gravelled yard at the back used for iron working. Clearly, the century that eventually built its barracks in that space had been away since the foundation of the fortress. Between the rampart and the *via sagularis* would have been a number of timber structures, buildings of a non-military nature, such as ovens, rubbish pits or even toilets. Outside the fortress, the civilian settlement came almost up to the walls. Work was also beginning to repair the amphitheatre, and scaffolding may have been over the baths as it too was refurbished after a long absence by the legion.

By AD 200, things would have changed considerably. The rampart towers, at least on this section of the fortress wall, had been removed to make way for other buildings. In the intervening 100 years, Barrack XII had been built in stone, demolished and replaced with a clay-surfaced hard standing. It would stay that way until the end of the century, when a hearth and a pit occupied the site. Cookhouses were now attached to the bases of the disused towers, which themselves were filled with ash and charcoal from fires. Next to the cookhouse was a stone latrine, and also cut into the rampart on the side facing the entrance to the barracks were a number of long buildings. These rampart buildings have been seen as extensions of the barracks, where legionaries could congregate and eat, but later were turned into arms stores where repairs might be carried out. The legionary on the rampart could look over the fortress and see barracks that had been rebuilt with their orientation changed from being parallel to each other, to facing across a courtyard – probably the decision of the camp prefect or cohort commander on return from a long absence when the structures were in a decayed state. Outside the fortress, the civilian settlement had been destroyed by a fire and rebuilt behind a wall away from the defences, this space was now used as a parade ground.

The one thing that would have been constant was a continuous supply of legionaries to replace those who had completed their twenty-five years or had died in service.

2

The *via praetoria*

It is rare to discover the names of serving legionaries in their barrack block, as it is usually only on tombstones (as veterans) that their identity is revealed. However, from Prysg Field Barrack VIII we have two cast-lead dies, probably used for stamping objects with the names of the two legionaries. Both Corellius Audax and Sentius Paullinus were in the century of Vibius Severus, and may have served in the legion *c.* AD 100. Unfortunately, we are unable to ascertain whether they were recruited at the same time and in the same place, or transferred from another legion, but since these objects were found in the same room it is entirely possible that they were mess mates.

Aristides writing in AD 143 commented about the recruitment process:

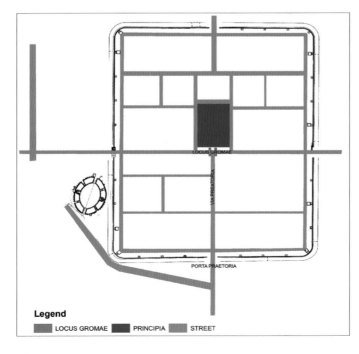

Legend
LOCUS GROMAE   PRINCIPIA   STREET

The *via praetorian.*
(*Caro McIntosh*)

When you had found them, you released them from their native land and gave them your own city in exchange ... On the day they join the army, they lost their original city, but from the very same day become fellow citizens of your city and its defenders.

Aristides (*Eis Romen* 26K, 74–77)

RECRUITMENT

Service in a peacetime legion was much sought after, and there was no need for conscription. A wealth of volunteers testifies that most legionary soldiers appear to have been content with the pay, which was more than they would have earned as civilians. There were opportunities of promotion through the ranks, theoretically to high positions in the legion and then, after retirement, in civilian life. The diet was varied and dependable, there were good health provisions and a variety of exciting experiences both inside and outside the fortress.

Initially, Italian recruits would have still predominated at Isca in the early Principate, but as the numbers of volunteers declined, about the time of Hadrian (AD 117–138), their place was taken by men from the veteran colonies of retired troops founded mainly in the provinces, especially in the areas of Gallia Narbonensis on the Mediterranean coast of France (Roman Gaul), Spain and North Africa. Each legion drew on its own on veteran colony and the sons of retired legionaries; in the case of the Second Augustan Legion it was at present-day Orange, Colonia Julia Firma Secundanorum Arausio (the Julian colony of Arausio, established by

Dies for stamping the objects made by Corellius Audax and Sentius Paullinus of the century of Vibius Severus. They were found in the *papilio* (the living room) of one of the *contubernia* of Barrack VIII of the Prysg Field series and dated by the excavator AD 75–105. They were made of lead with names being cast in retrograde and are 15 cm long, 5 cm wide and 0.8 cm thick. (*Archaeol Cambrensis*, 1932, Vol. 87, Part 1, p. 52.)

the soldiers of the Second Legion) and founded in 35 BC. In the first century, the sons of legionaries living outside fortresses in the *canabae* settlement began to be recruited, and increasingly so throughout the second century until they became the most important source in the third century. In Britain, if Roman legionaries were marrying local women and their sons were therefore Roman citizens who 'joined up', the legion became increasingly 'British' over time.

Recruitment was vital in keeping the legion up to strength as there were continual losses from retirement and, more rarely in Britannia, through military action. When men were needed, an office was opened up in the area of recruitment and, in the way of all permanent armies, each stage of the recruit's progress was recorded many times for his dossier. Each man had to demonstrate that he was eligible to join the army (*probatoria*), and the correct documents were needed to demonstrate that he was not just a freeman, but also a Roman citizen. The recruit also had to produce references to ascertain his suitability for the role (and, as usual, the more influential the referee the better). If all the documents were acceptable then a rigorous medical examination was undertaken. If the medical was passed, the governor of the province would approve the man for military service and the recruit would be *probates* (an officially enrolled soldier). All the information so far collected was contained in a record for the recruit's first post, accompanied by a letter from the governor. Vegetius is our main source for the qualities that were looked for in a legionary. He suggests that the recruit comes from a temperate climate, rather than a hot one, as he will have sufficient blood to inspire contempt for wounds and death, and also intelligence that maintains order in camp, and is of considerable advantage to plans in battle. Vegetius recommends men from the countryside are to be preferred to those from a city, as a life of hard work was more suitable than that of a luxurious one. The minimum height was six Roman feet (a Roman foot was 11.7 inches – 5 feet 9 inches), although he insists that bravery was more important than height. The correct age was the beginning of puberty, and studies of military careers from documents and gravestones indicate that although the range was from thirteen to thirty-six years, the majority of recruits were in the eighteen- to twenty-three-year-old group. The theory was that younger men would learn instructions more easily, and have faster reactions than older candidates. Other points to look for were lively eyes, carrying the head erect, a broad chest, muscular shoulders, strong arms, long fingers, a modest belly, thin buttocks, sinewy feet and calves, and not overly fat. Manly trades were preferred for potential legionaries, such as smith, blacksmith, wagon maker, victualler and huntsman. Vegetius insisted that the most important qualities were modesty, honesty and industry. If the recruit was suitable, each one was given *viaticum* (a travel allowance) for the journey to the chosen unit, possibly accompanied by the legionary doctor who had undertaken the medicals.

While there are no accounts of the arrival of recruits in a fortress, we can use the literary and archaeological information to 'transport' them to Isca. Let us put them at the top of the hill overlooking the fortress from the east, having travelled by straight roads from one of the ports on the southern coast of Britannia. The two buildings that would have stood out above the flat-roofed profile of the fortress were the huge cathedral-sized *thermae* (the bath complex) and the tall *principia* (headquarters), both of which said 'Roman' and were meant to emphasise the control of the landscape to anyone, both local and military. They would descend the hill, cross the river by bridge and enter the fortress (after being challenged by a guard) through the *porta praetoria* (main gate of the fortress) facing east, the recommended direction if there were no enemy in the area. The gateway had two passageways, which were surmounted by a wall walk, and probably an inscribed plaque announcing the name of the fortress and legion, or when the gate had been (re-) built, and by whom.

Once through the gate, and inside the fortress, the *via praetoria* stretched out in front of the recruits. Hyginus informed us earlier that the legionary standards in the *principia* 'should look down' on to the *via praetoria*, the converse being that anyone entering the fortress through the *porta praetoria*, would be looking 'up' at the façade of the *principia*. The lines of colonnades flanking the street would have hid the buildings behind, and led the recruits' sight line to the dominating three arches of the *groma* that acted as the monumental façade of the *principia*.

THE *GROMA*

As we have seen, Vegetius suggests that when the fortress was being laid out, the second act was 'to plant the ensigns, held by the soldiers in the highest veneration and respect, in their proper places', which might well be at the point of the *locus gromae*. It would not be out of place if, using a temporary altar, there was a religious ceremony conducted by the legate to confer a blessing by the gods on the legion's new fortress. The building later known as the *groma* might be a permanent representation of an altar with the appropriate inscription, commemorating the conquest of the local peoples, the Silures. Such a structure would hardly have been constructed in timber and must have been in stone from the beginning.

The governor of the province, from which they were recruited, had given the men a letter for the unit commander with a list of names, ages and a physical characteristic to identify each and the precise date from which he was *probates*. A copy of this information was also kept at the headquarters of the provincial army in which he was to serve. On the day of arrival at the unit, the recruit would have to present himself at the *tabularium legionis* (administrative office of the legion) sited at the back of the *principia*, where his joining the legion would be recorded *acta diurnal* and an individual record started. Walking through the sequence of the rooms of the *principia* would have been a daunting and humbling experience with a courtyard (*forum militare*), then a massive and magnificent hall (*basilica principiorum*), at the back of which was the temple of standards (*aedes principiorum*). Although we have no record, it would be entirely appropriate for recruits to make their oath to the Eagle of the Legion in front of the *aedes*.

The *groma* acting as a monumental entrance to the *principia* at the fortress of Lambaesis in the Roman province of Numidia, now Algeria. The large arch is where the *via principia* enters from the *sinistra* gate, and the two smaller arches either side are for the pavements of the street, with the one on the far right leading to the range of rooms inside the *principia*, as can be seen by the stub walls either side of it. The entrance on the left is for the *via praetoria*. The importance of the street is reflected in the decorated façade, with its Corinthian columns, *pilasters* and, on the key stones of the arches, sculpture, including a Victory, a genius and a legionary standard. It would have been this central arch that focused the recruits' eyes through the headquarters building to the *aedes* where the totems of the legion, particularly its god-eagle, were displayed. (*R.V. Schoder, Loyola University Chicago Digital Special Collections*)

TO THE CENTURIES

The recruits were then distributed around the centuries that needed men, and would be given four months of training, after which the results of assessments of each man's stamina and character were entered into his file in the *principia*. Again Vegetius gives us the details of what the training programme comprised. Each recruit learned the arts of weaponry, using special practice swords, shields, helmets (with a wooden post as the enemy), drill and movements, to never desert, to keep ranks, to throw their weapons with great force and accuracy, to dig ditches, to plant a palisade with skill, to handle a shield and deflect the oncoming weapons of the enemy by holding it at the right angle, to avoid a blow with skill, and deliver one with bravery. Other exacting tasks were undertaken, such as marching at the military and rapid paces, swimming, vaulting on horses, felling trees, and carrying heavy packs on long route marches three times a month.

If the recruit proved satisfactory at a proficiency test, he would attend a passing out parade when he would receive the *signaculum* (the military identity disc worn around the neck) and be *signatus*. He was entered on the records of the legion as a fully trained and qualified legionary to be treated as *milites* legally and receive a legionary's pay. It was then that he took the full military oath to the Emperor (*sacramentum militiae recusavit*).

As a fully qualified member of the legion, the recruit would join a *contubernium* of eight 'fellow citizens' of the legionary city, based in the barracks on the *via sagularis* that ran inside the defences around the fortress.

3

The *via sagularis*

The *via sagularis* was named from the red cloak (*sagum*) that completed the uniform of the individual legionary who assembled there for inspection each day, and so this street is intimately related to the common soldier, such as Corellius Audax and Sentius Paullinus.

Both belonged to the century of Vibius Severus, whose eighty men lived in a barrack block as members of a *contubernium* (group of eight legionaries), who formed the most basic group in the legion, and who, it might be conjectured, lined up together for inspection. One of the terms for comradeship within the Roman army was *contubernalis* (tent or mess mate), and although the individual *contubernium* had neither strategic value nor tactical significance, it was a major support for the legionary, as the loyalty of his fellow soldiers was crucial in battle or on exercise. In such a large military organisation, the *milites* would be fighting for those they knew, rather than the distant person of the emperor or an elite concept such as the *Res Publica* (Roman State).

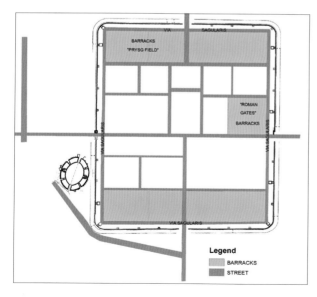

The *via sagularis* and the barracks of cohorts II–X.

Alan Sorrell's reconstruction of the early stone-built Prysg Field barracks using the excavator's report. On the right is the fortress wall with a sentry on the wall walk and men are lining up for inspection on the *via sagularis*. An *onager* (a stone-throwing machine) can be seen being moved. The barracks with their larger centurion's quarters have a veranda in front of the cubicles. Smoke is rising from the ovens at the back of the rampart. (*Reproduced with permission of Julia Sorrell*)

BUILDING THE BARRACKS

Hyginus tells us that the tented barracks were to be distributed around the fortress perimeter along the *via sagularis* for quick access to the rampart walk, so that troops could be moved quickly to any part of the camp defences in case of attack. He also tells us that on campaign, the tents of the *contubernia* had to be in a straight line behind the centurion's double tent, and that each line of tents had to be parallel to the other ten lines of the whole cohort (this was written before cohorts were reduced to six centuries and the century was changed to eighty men). At all legionary fortresses, the tented arrangements transmuted into timber barracks, and at permanent bases were later built from stone on the same footprint. At Isca and other fortresses, this resulted in each cohort being housed along the *via sagularis*, except for the prestigious first cohort.

At the Prysg Field blocks were paired across the courtyard, making a *strigae*. However, the barracks at the Roman Gates site were in a line with no shared space,

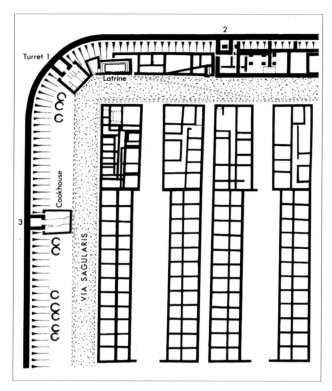

Left: The plan of the Prysg Field excavations of 1927–29. (*after Archaeol Cambrensis 1931, Vol. 86, Part 1*)

Below: The view from the rampart with the *intervallum* containing the early ovens, and the drain alongside the *via sagularis* with the barrack block behind. One of the cookhouses can be seen in the middle distance. The difference in height between the rows of barracks is the result of the nearest being completely uncovered, while the others were only trenched and their 'footprints' are laid out at modern ground level. (*Author's collection*)

and this may have been the choice of the cohort commander when they were originally constructed. During a rebuild later in their history, these barracks were made to face each other, which created a courtyard, again possibly a choice of the incumbent cohort commander. The layout and the planning of the barracks in the *insula* allocated to them may have been negotiated by the centurions, and then laid out by the pacing, usually either side of 50 paces – the average at Caerleon Prysg Field being 48.03 paces. At Isca, the barracks at the rear of the fortress, including the Prysg Field, were noticeably shorter than the others known in the base, with only twelve *contubernia* rather than thirteen elsewhere. This may indicate the lower status of the cohorts. At Inchtuthil, Shirley calculated that a barrack block XVII, which was a little longer than those in the Prsyg Field, would have taken somewhere in the region of 250 man hours for excavation and backfilling, 1,450 for wall framing, and 2,000 for roof framework – a total of 3,700 man hours. The most efficient way of constructing the block was for the *milites* to provide the labour, and the *immunes* (specialists in timber construction) supervised them. Assuming an eight-hour day, a workforce of forty men (half of a century) could have built the block in about twenty-three days, not including the supply of materials and equipment, which may have lengthened the build time. Rebuilding the barrack in stone would have taken much longer, with the foundations being strengthened and fired clay tiles fitted over the whole block. It is possible that the structure was prefabricated elsewhere, and moved onto its permanent site. The materials for construction of the blocks would have had to be sourced from outside the fortress, possibly by an advance party, but may also have been transported from Exeter, the last fortress occupied by the legion. Where the timber phase was seen at the Roman Gates, it was of post-in-trench construction, a technique used to delay rotting – a ditch was dug, a wooden cill put in it and the posts slotted in, forming the uprights of the framework. This is a different method from post-pit construction, where the upright was driven into prepared ground with flat stone at the bottom of the pit. Whichever technique was used, the superstructure would have had wall frames of wattle and daub, comprising of woven branches and mud possibly mixed with horse hair. It is most likely that the partitions between the rooms were wooden. The walls of the stone phase were usually constructed of roughly dressed blocks set in clay rubble core, often with the first two courses underground. There have been discussions about whether wattle and daub was used to form the walls in the stone phase. However, at the Roman Gates, the foundations could easily have taken a single-storey stone building, and in one case a double-storey. If the walls were of sandstone there could be rising damp, which would gradually weaken the rock and call for frequent rebuilding.

The foundations of a barrack block can be seen in the Prysg Field at Isca, as the walls above have either been robbed for stone or destroyed when the building was made redundant. Once the barrack blocks had been constructed, these foundations would have been invisible to the inhabitants.

Barrack Block I with the centurion's quarters closest to the viewer, highlighting the difference in space compared with the *contubernia*. The main drain carrying effluent from the toilets is running alongside the barrack block, and the *via sagularis* has turned through ninety degrees to follow the defences. (*Author's collection*)

THE *CONTUBERNIA*

The ground plan of the barracks emphasise the huge social and financial differences between the centurion and his century. The centurion's quarters took up 30–40 per cent of the length of a barrack, and might be detached and separated from the main block by a narrow alleyway. There is evidence of contrasting roofing materials; the senior officer had tile while the rest was covered in wooden shingles (though it seems not at Isca). The centurion's pay in the second century was ten times that of the individual *miles*, and much more than his officers, and so he had the finances and freedom to design and equip his quarters as he wished. The legionaries had virtually no way of imprinting their identities on the small space available to them in their ten *contubernia*, which were made up two rooms – the *papilio* for everyday living up, which took up two-thirds of the width of the block, and the *arma* for equipment. During the active life of the fortress, the barracks would have been rebuilt or repaired several times, yet the only element of the plan to change was the centurion's quarters, which probably reflected the taste of the individual officer. The *contubernia* retained their simple footprint.

The recruits, such as Corellius Audax and Sentius Paullinus, may have replaced either a retired legionary or, less likely, one killed in action, and it must have been difficult to breech the comradeship of the others in the *contubernium*, especially if the veteran had opted to live within the proximity of the fortress to be close to his former

The *contuburnia* with the *papiliones* nearest the road with the *armae* beyond. The plan of these rooms remained fixed throughout the occupation of the fortress. (*Author's collection*)

mess mates. From the earlier Neronian fortress at Usk, Burrium, graffiti on fragments of a mixing bowl (*mortarium*) where found in adjacent pits. The lettering was cut before firing and inverted so it could be read from above, and stated that it belonged to the *contubernium* of Messor. The inscription suggests that each set of barrack rooms had a leader appointed from among its members. This legionary, possibly the one with the longest service, may have been responsible for the behaviour of the other seven men and the cleanliness of the two rooms. No doubt a Messor would have eased the recruits into the life of the group, especially as the new members had been through a hard regime of training, which would have engendered respect, and would go through similar drills each day with their fellow soldiers. Bonds were also strengthened not just through training together, but through sharing daily duties, at leisure and mealtimes, either in their room or in close proximity to the barracks. These close ties between the eight men also had to be strong in order to share a cramped room together while in barracks or a tent on campaign. Corellius Audax and Sentius Paullinus in barrack block VIII probably had about 2.8 square metres of space each.

The eight men in the *contubernium* would have been *milites* (infantrymen) and possibly *immunes*, who were exempt (immune) from the more tedious and dangerous tasks of ditch digging and rampart patrol, and who were on the books of the centuries, and presumably were quartered with them. They were paid at the same rate as the *milites*, but had to complete at least three years as an infantryman before specialist training for a specific role. It is most likely that they supervised work being undertaken by the *milites* and, considering the position of the Prysg Field barracks, it is likely to

Another view showing the difference in space allotted to the centurion and the individual *contubernium*. The remains of the veranda are closest to the viewer. (*Author's collection*)

have been in the *fabricae* (the workshops). A conservative estimate is that between seventy and 150 soldiers in the legion were *immunes* and worked in most parts of the fortress: in the hospital, the workshops and clerks to the higher ranks in the hierarchy of the fortress, as well as musicians.

Since the Prysg Field explorations of the late 1920s, there has not been a set of barrack blocks excavated in their entirety, although parts of the centurion's quarters have been explored in York, Colchester, and Chester. An exploration on the *contubernia* at the Roman Gates in 1980 has produced a large range of finds that have given us a comprehensive view of the everyday life of a legionary. An exploration of part of the previously unexcavated barracks of the Prysg Field series at Alstone Cottage in 1970, and at 'Sandygate' in 1985, gave valuable insights into both the centurion's and men's quarters respectively. Clearly, there will be a fuller picture of life in the barracks if the results of these all these excavations are taken together.

CONDITIONS IN THE *CONTUBERNIA*

The lighting in each of the cubicles would have been a problem, especially so in the barracks if the long side was back-to-back with the next, blocking the other's light. The quality of light in the *arma* would have been lessened by the existence of a veranda, which, with the partition wall between the rooms, would have the affected the *papilio* even more. Considering the amounts of window glass found in all barrack excavations, it is clear that some, if not all, of the *contubernia* had glazed windows, although these might be high up in the walls. There is enough evidence to show that the use of painted plaster and whitewash also might have been used to make the cubicles lighter. However, the large number of lamps, either pottery or lead, would support the notion of limited daylight in the *papiliones*. The distribution of the origins of the lamps (some of which had come from some distance, such as the Rhineland, Gaul and Italy) is interesting in that it reflects, to some extent, the origins of the legionaries and might have been curated over long periods as essential items, or even received from family. Candleholders are also a frequent find.

The floors of the rooms were of clay, concrete or sometimes flagstones. Clay would have not have been seen as a low status material, but more appropriate for the British climate, being less hard and cold to walk on than stone, especially considering the problems of damp. In barracks elsewhere, hearths were employed for heat and cooking, but at Isca there is almost a total absence of these features. At Alstone Cottage, patches of pebble floor and areas of charcoal in the *papiliones* and continuous layers of charcoal in *armae* suggest braziers, however, no remains have been found. A position for ovens against the partition between *papiliones* and *armae* would be under the highest point of the roof, and well-positioned for a chimney or smoke outlet. These would have been dangerous in the timber phase of the barracks, however, so perhaps there were different cooking and heating arrangements between the wooden and stone phases of the structures. Many excavated barrack blocks appear to be full of rubbish, but this did not necessarily accumulate while the blocks were in use, as we have documentary evidence that one of the duties that could be assigned to a legionary was to sweep the barrack floor. The blocks might have acted as dumps if a century was absent from its cohort for some reason. On return, one way of dealing with the situation would have been to put a new clay floor down to cover the refuse, or even make a timber floor to go over it. No doubt the wooden, wattle and daub partition walls would have increased noise levels throughout the men's part of the block. Overall, it is difficult not to come to the conclusion that the *contubernia* were often confined, rather basic and sometimes squalid.

We know so little about what happened under the veranda, even if it usually existed, as they have not been regarded as spaces deemed worthy of complete excavation, although there is an example of a paved external floor at Barrack I of the Prysg series. Storage pits and latrine pits lined with wickerwork have been found in verandas in some fortresses, which would make it uncomfortable in the cubicles next to them. On sunny days, the clay street between the barracks facing each other would have acted as a suntrap, making conditions even worse. No doubt there would have been friendly banter and competition between the centuries across the courtyard, or the space might have been a perfect arena to settle a personal grudge.

BARRACK LIFE

While we have Vegetius to tell us about strategies, the layout of the camp and recruitment, we have little information about the ordinary soldier, and we must turn to archaeology to overcome the problems of understanding barrack life. However, the archaeological resources available to us depend on the materials of which the objects were made from, and if they have survived. Owing to this issue, we still have no evidence of many aspects of life, such as sleeping arrangements. It is often presumed that bunks were used, though it would have been a squeeze to fit in eight men. Considering that some men would be on duty each evening, or perhaps on leave, the use of 'hot bedding' is possible, but so is the employment of straw mattresses, which could be tidied away each day. Hyginus, when writing about a temporary campaign camp, suggests that not more than eight tents are pitched, as sixteen men are always on guard duty at any one time.

MILITARY EQUIPMENT

Sleeping arrangements did not concern the Emperor Severus Alexander (AD 222–235), who is said to have had the maxim, 'One need not fear a soldier, if he is fully armed, properly clothed, has a stout pair of boots, a full belly and something in his money belt' (he was murdered by the army!) The legionary had to buy his own armour in the form of pay deductions, and it was in his interest to ensure that it was in working order and fit for purpose. This is probably the main reason that large items of military equipment are not found in barrack blocks, but only small and relatively insignificant objects. The majority of the finds are small bronze fittings from the *lorica segmentata* (armour) worn by the legionaries.

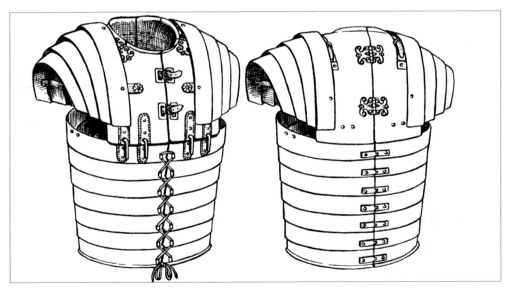

Lorica Segmentata (the name in Latin is probably sixteenth-century, and the Romans might have referred to it *lorica laminata*) armour with a large number of fittings that could detach themselves. Add to this the hangings from leather aprons worn underneath and studs from helmets, there is a rich hoard of metal to be found in barrack blocks. (*Author's collection*)

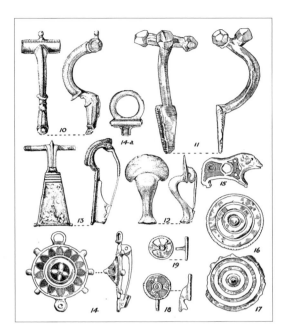

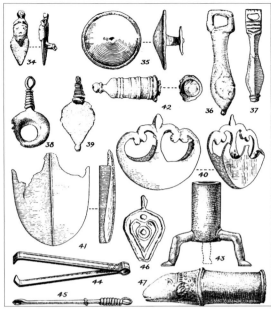

Above left: The Prysg Field excavations produced a wide range of metal finds relating to the daily lives of the *milites*. Nos 10–14 are brooches, perhaps worn with the *sagum*. No. 14a is an eyelet from the *lorica* (armour) and was common in the *contubernia* where they were easily lost in the darkness. Objects 16–18 are also brooches, but more highly decorated and show either some personal variation of uniform or were for less formal occasions. (*Archaeol Cambrensis 1932, Vol. 86, Part 1, p. 82*)

Above right: Nos 34–39 are bronze mounts connected with leather fixings, while 40 and 41 are the capes for sword sheaths, and 42 the ferrule of a *pilum*. Items such as the tweezers (44) and the pin (45) are often connected with personal cleanliness, although are often connected with hospital instruments. The candelabra (43) and possibly the head of saucepan (47) give a flavour of everyday life and personal possessions of the *contubernia*. (*Archaeol Cambrensis, 1932, Vol. 86, Part 1, p. 86*)

The complexity of this type of armour was shown in the loss of tie rings, buckles, buttons, hinges and studs for joining segments of armour, belt plates, strap ends from belts, pendants that hung from the armour rings and from chain mail, as well as apron fittings. Helmet fittings are also common – suspension rings for fastening them, crest supports and a plumb tube for the crest itself. In many cases, the individual pieces did not match, and had probably been repaired by the local armourer. Personal weapons are evidenced by the individual designs of scabbard fittings, bindings, chapes and bone handles that had been lost or broken from the *gladius* (short stabbing sword). Ferrules from the bottoms of the *pilum* (javelin) are widespread, as are loop fasteners to aid throwing. Most of these items were small and, in the darkness of the individual *arma*, their loss would not have been noticed and were easily replaced. This is obviously the reason why so many hobnails from boots are found. Whole items such as personal

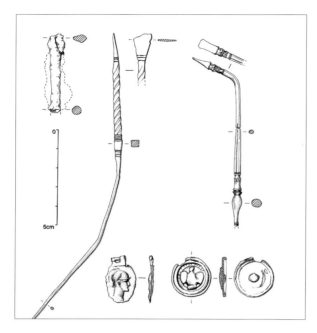

Styli and seal boxes, one with an eagle, from the Blake Street barracks in York. This might be evidence of *immunes* working as administrators. (*York Archaeological Trust*)

daggers and spearheads are not common in *contubernia*, which is understandable since the legionary would take his personal weapons with him if troop movement occurred. Considering the high water table at Caerleon, it is surprising that more perishable items haven't been found, but even if the woollen clothes did not survive, we have the evidence of a variety of brooches for holding military and other cloaks.

The veranda at Alstone Cottage had an unusual stone cist, which was 0.60 metres by 0.48 metres by 0.38 metres deep. It was floored with a single flat stone 0.33 metres by 0.38 metres by 0.18 metres thick, full of brick dust produced by the pounding of tile fragments; the purpose was certainly to produce stone material for cleaning metal equipment, especially the rust prone *lorica segmentata*, (the Romans probably referred to it as *lorica laminata*) and edged weapons. Vegetius comments that it is the centurion's duty to ensure that the men have 'their arms constantly rubbed and bright'. That is certainly why so many whet stones for sharpening weapons are found in the men's rooms of the barracks.

Evidence of the specialist roles within the fortress carried out by the *immunes* were found at Roman Gates, and at the Blake Street barracks in York, including a bronze inkwell and an iron stylus, which indicates some sort of literacy, and survey equipment in the forms of plumb bobs and weights. Items that might be used in the hospital, such as probes and spatulas, were also retrieved, and, since these barracks are close to the hospital, might indicate that this was the main accommodation for the *immunes* who worked there. Indications of horse specialists are suggested by the finding of cart hub lining and fragments of pins. Small bells might have been part of horses' harnesses, or as part of *tintinnabula* to ward off evil spirits and hung in the breeze outside the barracks.

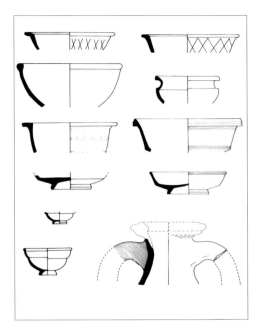 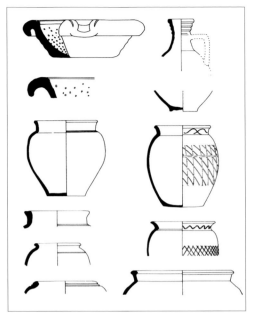

Above left: A selection of dishes, cups and an *amphora* for wine, olive oil or fish sauce from the Northgate Brewery barracks in Chester. (*Copyright of Cheshire West and Chester Council*)

Above right: A selection of storage and cooking pots, as well as a jug spout. (*Copyright of Cheshire West and Chester Council*)

COOKING AND EATING

The Roman army had no general messes for its soldiers, and no dining halls in its fortresses, although there is evidence of a *taberna* on the *via principalis,* which seemingly sold chicken and wine – a 'fast food' outlet. The eight men were expected to prepare their own meals and to pay for the food, which was deducted from their wages. As well as the pleasure of eating together, group identity was also fostered by discussing the day's duties and probably being irritable about the unpleasantness of duties, or the centurion and his deputy (*optio*). Feeding eight men would require a series of vessels of considerable size, some necessarily of iron or some other metal. Finds of such vessels, and means of support over braziers, are almost unknown at Isca, yet not all could have rusted away. This leaves the possibility that the men ate outside the individual *contubernium* and, as Messor's pot indicates, each had its own crockery and mess utensils. In each excavated block, there are a range of pottery jars, flagons, wide-mouthed bowls, cooking pots and mixing bowls, known as *mortaria*, as well as glass bowls jugs, beakers and jars, which demonstrated preparation and storage of food.

Examples of quern stones and hand mills for grinding wheat and oats into flour for loaves or biscuits are common, but they were probably also used for pulses that were crushed for porridge, soups or pottage. In terms of a meat diet, ox, pig, sheep, goat, red deer and horse bones have been most common in barracks across the fortress. Finds of hooks suggest fishing in the local rivers in the legionaries' spare time.

Fragments of iron buckets to carry water from the nearest water trough, and pan and skillet fragments, give some idea of the cooking process. Knives, knife handles, spoons, handles from a bronze tankard and a bronze jug, demonstrate that legionaries could afford quality vessels. The expensive samian ware, the glossy, decorated, red pottery from Gaul is very common on legionary sites, and demonstrates the spending power of the common legionary. Scenes of hunting, gods and myths, gladiators, fighting and erotic scenes may well have been produced with the army as targeted consumers.

PEOPLE

The reasons that the country boys joined the army was the pay and the reward of money or land at the end of service. After being docked the money for his armour and food, contributions to his savings and burial club, the legionary had enough money to be spent outside the fortress in the *canabae* on personal weapons or on 'wine, women and song'. There were also opportunities for spending money when on leave. Excavated *contubernia* are always full of coinage, dropped and not recovered because of the darkness, or even hoarded until return from an expedition, as in one of the barracks in the Prysg Field. Pay was certainly spent on personal items attached to the individual identities of the barrack population. Bracelets, rings, earrings, glass-headed pins in bronze, hair pins in bone, melon-shaped beads and large numbers of armlets all point to personal decoration when out of full uniform. Many soldiers had rings with *intaglios* (carved stones) set in them, often with their favourite god for protection. A frequently used stone was cornelian, easy to cut and very popular among soldiers over the whole Empire. The semi-precious stones often fell out of their settings because the metal expanded when heated in some way, but the stone stayed cool. The number of locks and keys may have a relevance to these items being carefully stored in chests. All soldiers get bored when not on duty, or had little to spend their money on, so gambling was a favourite pastime, as evidenced by the finds of bone dice, and stone, glass and bone counters used in games.

Surprisingly, there has been very little evidence of religious associations in any of the barracks. Soldiers would have used the temples in the *canabae* outside the fortress to satisfy their own particular allegiance to a favourite god or goddess, in addition to the all-embracing, statutory devotion of the 'official' gods of the Empire and legion. If there were any personal devotional objects, they were undoubtedly taken care of by the individual, and taken with them on transfer or expeditions. It would be surprising if there hadn't been a protecting spirit for the *contubernium*, as well as some sort of daily ritual associated with it.

CENTURIONS

We have the names of many centurions at Isca, but most of them cannot be connected to specific barracks blocks, although from the Prysg Field barracks (AD 100–200) we know of Vibius Severus (the centurion of Corellius Audax and Sentius Paullinus), Vibius Proculus and Quintinius Aquila. We are still unable to surmise how long these officers were in command of the century, or even if they were contemporary. The centurions in a legion were responsible for the administration, conveyance of orders, the training of troops and leadership into battle. The maintenance of discipline was also major responsibility; the vine stick (*vitis*) was an essential badge of his rank, as was the centurion wearing his sword (*gladius*) on the left and his dagger (*pugio*) on the right, the reverse of the legionary soldier. Promotion to the centurionate was usually the gift of the provincial governor, who was the nominal commander in a province, but there was also a mechanism whereby the legion could vote for an individual to be promoted to the rank. There were four main routes to becoming a centurion. The most common was probably up the ranks from an *optio*, the century's second-in-command. Another possibility was serving six years in the Praetorian Guard, the urban troops of Rome, and one of the results of this would be knowledge of the area around the capital and the upper-class culture, which might be reflected in the way individual centurions displayed their wealth in their barrack block. Centurions of auxiliary units would also have qualified for centurial status in the legion, and this would have been a promotion (legionary centurions might also be in command of an auxiliary fort). In each of these cases, the man would have already passed the recruitment requirements and have been well trained in army matters. However, there was also a direct commission route from the equestrian order, but centurions coming through this route would not have had much professional experience of soldiering, and would have relied heavily on their *optio*. We have no knowledge of what fraction of the centurions came through each route, though it is likely that the most common was up through the ranks.

From records in other parts of the Empire, we have evidence that direct commissions were probably thirty to thirty-five years old, and those coming up from the ranks were probably of the same age, having already served around fifteen years. There seems to have been no set period of service, with some careers ending in death. Petronius Fortunatus served for fifty years with thirteen legions, including the *Legio II Augusta*. He had climbed through ranks, from clerk to officer in charge of the watch word, standard bearer, *optio* and then centurion. His son, Marcus Petronius Fortunatus, was also a centurion who served with the legion, but died aged thirty-five, having served six years in the army. It would seem that he came in by a route other than promotion through the ranks like his father. A legionary centurion was chosen for his size, strength and dexterity in throwing his missile weapons, and for his skill in the use of his sword and shield; in short for his expertise in all the exercises.

He is to be vigilant, temperate, active and readier to execute the orders he receives than to talk; strict in exercising and keeping up proper discipline among his soldiers, in obliging them to appear clean and well-dressed and to have their arms constantly rubbed and bright.

                                                                Vegetius

It seems that centurions were not to have initiative, or take decisions in battle.

How the centuries of cohorts II–X were ranked is still unknown, although it has been suggested that those at the rear of the fortress, the furthest away from the main entrance at the *porta praetoria*, were higher numbered and of lower status. The barrack blocks in this area of the fortress do seem shorter, but that might be for reasons other than prestige. We also can't be certain of the ranking of centurions in the individual cohorts, as there is no evidence that the size of the centurion's quarters reflected standing, and it is most likely that the officers were of equal status, but listed by seniority of service. This may be the explanation why at Isca, Roesius Moderatus was a junior centurion in the VI cohort. Each cohort had six centuries, here listed from the most junior:

Hastatus posterior
Hastatus prior
Princeps posterior
Princeps prior
Pilus posterior
Pilus prior (cohort commander)

At the Prysg Field, the barracks that were excavated in the 1920s and '30s can give some ideas of the manner in which the status of the centurion was displayed, and many of these aspects were confirmed by the excavations at Alstone Cottage, Barrack XII. In 1970, the two end rooms of the centurion's quarters and two of the *contubernia* on the 'centurion wall' were sampled. Although not confirmed by excavation, one would expect a stone wall between the centurion's quarters and the *contubernia* to lessen noise from the legionaries. This may well be the reason why, in some fortresses, the centurion's quarters were detached from the rest of the block.

The design of the centurion's quarters was simply a corridor with rooms off of it, and it is likely that the layout was the personal preference of the officer who was in charge when the building was first constructed or rebuilt in stone at a later date. Even in the stone phase, the centurion's quarters were often rearranged for personal choice. The external door, as in all the barrack blocks, was on the short wall facing the *via sagularis*, and the entrance in some was paved, no doubt to protect against heavy wear and the effects of wet boots coming in from the road. The rooms at the end of the block near the *via sagularis* all seem to have some sort of washing/toilet function, often with the water passing into the drains alongside the road. It may be the drains were an outlet for sewage, but since there is no evidence of piped water, buckets would have to be used for flushing away the results of the activity. In barracks where there was no outlet, water containers made of stone have been found, which may have been used for cleaning and washing. A bath was found in these rooms in the centurion's quarters, in the north-eastern corner of the fortress.

Above: The washing facilities emptying into the main drain in the front portion of the centurion's quarters. (*Author's collection*)

Right: The corridor of the centurion's quarters in the two stone phases revealed by the excavation. (*Author's collection*)

Usually, the accommodation in the middle of the centurion's quarters was floored with clay, although it is possible that wooden floors were suspended above them. Wood would be much more effective in areas where water was used and certainly would have given a comfortable feeling to other rooms. These middle rooms were also used for storage, or as a kitchen, as a concentration of course-ware containers (*mortaria*) tools and weapons have been found in this situation in Chester. The rooms at the end of the corridor were furthest from the entrance, where there was less noise from the road, and few cold draughts. These spaces were also the furthest possible distance away from the latrine and sewage running along the *via sagularis*. There is no doubt that these were the centurion's private quarters, which had concrete floors, possibly with a good deal of broken tile embedded in them, making it an attractive reddish surface, likely covered by some sort of insulating material, such as rugs. Fragments of a mosaic dumped in the rampart buildings opposite the barracks at Isca may also have come from the rooms of an officer of this rank.

The more recently excavated centurial rooms of Barrack XII give an even more detailed picture. In the room overlooking the veranda, there was evidence of window glass and possibly two window openings, one onto the veranda and the other the courtyard. The floor was made of finely surfaced concrete, and the south wall of the room, backing on to the *contubernia*, had polychrome paintings consisting of a scheme of foliage rendered in light- and dark-olive green with red and yellow flowers with possible birds in dark brown. This scene seems to have appeared above painted panels of yellow and red. The centurion responsible for this scheme seems to have had personal experience of opulent living with an Italian background, and it is also likely that an outside expert would have to have been brought in to execute such a fine piece of work. Such grandeur could also represent the aspirations of a centurion who has risen through the ranks. This decoration may have been present in rooms in other centurion's quarters but, due to destruction and the trenching excavation techniques, these might have been missed. What is clear is that there is a firm correlation between the plans of centurions' accommodation in a fortress, and those of Italian apartments and flats as seen in Pompeii and Herculaneum, but without the reception rooms, which would not have been needed.

The rooms in the centurion's quarters were kept clean, and therefore finds are few. However, in other fortresses, high-class tableware (silver for senior commanders, and copper alloy vessels for centurions) and jewellery have been found, but, being portable, they were probably taken away from the fortress on campaign, transfer or retirement more often.

PRINCIPALES

This brings us to the accommodation for the equivalent of non-commissioned officers. It was once thought that they shared the centurion's quarters, however, the social and financial gradient between the two levels of rank was large, with the *optio* (the centurion's deputy) and *signifer* (standard bearer) only getting twice the legionaries' pay, and the *tesserarius* (guard commander) one and a half times. It is known that in cases of munity, the *principales* sided with the men, and vice versa, against the centurion, so we might expect these officers to have rooms somewhere in the *contubernia*. There is evidence in some of the blocks that the first two *contubernia* closest to the centurion's quarters were slightly larger than the others. Could this be where the *principales* were accommodated, producing noise insulation between the men and centurion, and giving the junior officers status by being at the head of the *contubernia*? It would explain why twelve *contubernia* were built, but only ten needed.

THE DAILY PROGRAMME

We can identify some of the activities of the day through literary sources and use archaeology to add the detail.

The Night Watch

Hyginus tells us that while in a marching camp on campaign, 20 per cent of the legion would be on night guard duty. We have no idea of the length of the watch, and how many *milites* would have been involved, although the watches would be signalled by a trumpet. At that point, the person doing the rounds would check that all was well with the changing of the guard, and record it on a wooden tablet. Polybius informs us that a 'guard' is made up of four men, and in a fortress with a more developed plan and a civilian settlement outside, we would expect a higher proportion, so perhaps 25 per cent at any one time – twenty men per century. From duty rosters from an auxiliary fort in Africa at Dura – the *Legio III Cyrenaica* in Alexandria – and Polybius' account, we know that the *praetorium*, *principia* and the houses of the tribunes were allotted guards, as were the hospital, granaries, weapon stores and the streets of the camp. Besides the walking of the rampart between the thirty towers, there was also the patrol path on the outside lip of the ditch. Presumably, these points were the responsibility of members of the century, or the cohort alongside them. There were the four gates, with two towers on each, which Polybius tells us had ten guards each. Outside of the camp there were probably sentries in the civilian settlement and certainly on the quays by the river. Centuries working in the *prata* (the land owned by the fortress in the area), were probably undertaking similar duties around important buildings. Presumably, if much of the legion was away, guarding the perimeter would have been the major task of the caretaking garrison, as many of the structures inside the fortress would have been mothballed.

Morning Parade

As we saw earlier, the *via sagularis* got its name from the cloak, the *sagum*, which completed the uniform of each of the legionaries who assembled there for inspection each day. The *admissa* (the camp prefect's orders for the day) were read out, along with any letters from the governor of the province and news. A new watchword for the day was also shared and the centurion or his *optio* detailed the duties to be undertaken by the *milites*, and the *immunes* were already attached to a facility where their specialist skills were most in demand. Then all the legionaries reaffirmed their oath of obedience to the emperor. The allocation of duties at the morning parade involved tasks that were physically demanding and guaranteed to occupy the legionary for much of the day, allaying Vegetius' fears of mutiny. Each of these activities would have been recorded by the *libraria* (clerks) of the various officers within the legion.

Daily Training: the Parade Ground (*Campus*)

It is likely, although there is no specific documentary evidence, that one shared experience for *milites* and *immunes* was the daily training regime on the parade ground. Vegetius emphasises the army undertook weapon exercises each morning on the parade ground (*ludus*), the legionary amphitheatre, or if it rained or snowed, in the *basilica thermarum* (drill hall). At Isca, the parade ground was reached along the *via sagularis* and out of the fortress through the *porta principia dextra*. We are dependent on documentary evidence for the activities undertaken on the *campus*, and these will be explored below.

Other Tasks

As well as the important role of sentry duty to protect important areas of the fortress, the duty roster for thirty-one men of *Legio III Cyrenaica* for the 1–10 October AD 87 gives a range of jobs for *milites* associated with the barracks and the *via sagularis*, from sweeper in the barracks, to cleaning latrines and fetching water.

LATRINES

There appears to have been a very limited provision of latrines at Isca – one per two cohorts at each angle of the fortress. Excavations in the Prysg Field identified the main culvert carrying water through the fortress from an aqueduct. A side branch fed the latrines, and the resulting effluent was taken away by a covered gutter, which reconnected with the stone slab-covered main culvert running between the long side of a barrack block and the *via sagularis*. It is not possible to construct the route of the water that was used for drinking and washing, because any provision of lead piping would have been reused, stolen or removed once the fortress had been abandoned.

One latrine for two cohorts is difficult to explain, even if it was expected that a number of the centuries, perhaps a whole cohort, would be absent from the fortress at any one time. It is feasible that 480 men could use the one set of latrines, depending on how many seats there were and how long the men took. Another possibility is that the latrines were used only by centurions and the *principales* – there is no convincing evidence of toilet facilities in the centurion's end of a block. There is another possibility that the urine and faeces were being used in many of the industrial processes, notably tanning, taking place in, or more likely outside, the fortress. Could it be that there

The latrines with the drain for the effluent had a gulley for washing sponges on its lip. Behind is a later cookhouse backing on to a corner tower. The flat surface has been interpreted as the base of an oven. (*Author's collection*)

were piss pots on every corner for every *contubernium* or on the barrack veranda? In the baths at Exeter, the bases of *amphora* (jars for carrying liquids) were used for just this purpose. Could it be that the legionaries' faeces were also collected in some way? The Roman army stressed the need for healthy troops and slit trenches covered with wooden seats on the veranda of the barracks or in-between would not be consistent with this policy, even if the sewage was covered with a layer of soil. Perhaps the area around the parade ground where soldiers would congregate when off duty might be the site for blocks of latrines.

It is difficult to say if the seating of the latrines within the fortress would have been made of wood or stone slabs, as the latter would have been removed on the abandonment of the fortress, but there was a groove for water to clean the sponges, and a trough for cleaning hands. Perhaps part of the duties of the cleaner was to wash the sponges and sticks, or to replace them each day.

OVENS

Polybius, writing about the Republican army in *c.* 140 BC, gives us a good excuse for number crunching about the supply of, and storage requirements for, wheat, but it is so hemmed in by provisos as to be only the roughest of guides. He suggests that the daily ration of corn for each legionary would have been the equivalent of 0.94 kg. We do not know whether corn was allocated daily or monthly, but if daily the eight mess mates would have had 7.52 kg, which could easily be carried by one man. The daily corn ration for the century would be 75.2 kg, which might have taken three or four men to carry, depending on how it was contained. Producing flour suitable for bread making needs a considerable amount of preparation time, and it is unlikely that a large group of men, such as the eighty of a century, would either wish, or be permitted, to spend the amount of time necessary to produce flour for baking bread individually. A legionary producing a loaf for himself, or even his *contubernium*, with the type of individual quern found in the barrack blocks, would take the same amount of time as a communal mill producing bread flour for a large number. We have evidence for this from a stamp belonging to a century, which was impressed on an unbaked loaf. This was probably necessary if a number of loaves from several centuries were in the communal ovens in the *intervallum* between the ramparts and the *via sagularis*, and later behind the watch towers.

It is usually thought that the domes of ovens were constructed of clay, as illustrated in excavated towns in Italy, such as Pompeii. However, no trace of clay has been found at Isca and the upper structure may have been built of thin slabs of sandstone, corbelled into a roof and covered in earth. The door of the oven opened on to a well-constructed hob in a chamber passage, and when in use a fire would have been built inside the dome, and the food, presumably bread, cooked in the residual heat of the ash once the entrance was sealed. The cold ash was raked out into a pit dug for the purpose, and then covered with earth, sand or burnt gravel. These ovens must have been covered with some sort of structure to keep fuel and bread dry, but no definitive evidence has been located at Caerleon. Often it is thought that ovens were placed below the rampart in the *intervallum* for reasons of safety in a timber fortress, but there were fires in the barracks and in the workshops. It is more likely the location of the ovens was due to these flimsy structures being able to be easily pulled down in case of an emergency when they might be at risk from enemy fire, or block access to the rampart. When the ovens below the rampart went out of use at Isca, it would seem that cookhouses were built on the back of the corner and in interval towers (at this point the wall walk may have gone out of use). While some of these square structures do have evidence of layers of charcoal, it is not certain that it was from cooking rather than the structure being cut into an area of previous ash pits. In one case at least, there was no evidence of any baking activities.

It is possible that while part of the corn ration allocated to the men was taken for baking in communal ovens, the rest was used by the individual soldier for his own ends. While bread making is a relatively specialised set of tasks, the preparation of a simple cereal food such as porridge, or the *bucellatum* (hard biscuit), which could be taken with the soldier when working outside the fortress, was quicker, involved much less effort and could be made in the *contubernium*. Small querns for grinding flour are often found, some made from basalt from the volcanic area of Mayen, Germany. They were probably imported as 'blanks', with the disks then being shaped closer to the fortress. However it was used, getting the grain would have involved going to the *horrae* (granaries), which also lay alongside the *via sagularis*, and will be visited later.

ARMAMENTARIUM

The duty roster of *Legio III Cyrenaica* for the 1–10 October AD 87 lists men working in the armoury of a century as a daily task, and we know the post of *armorum custos* was filled by one or two men in each century. These roles appear to have been concerned with making sure that the men had enough of the required equipment, and this would have involved supervising the scrapping and repair of damaged items, selling kit to new recruits, and buying it back from veterans at the end of service for a sum of money (since its cost had been deducted from the legionaries' pay). These activities may have taken place in a number of long, solidly built structures behind the rampart and alongside the *via sagularis* in the Prysg Field. It has been suggested that they were used for administration, stores or mess rooms for the century, but a more convincing function is the storage and repair of personal and communal weapons. Stratified finds from the earliest phase of the structure produced seven scabbard fittings, nine chapes from swords and four spearheads, as well as a copper harness roundel for a horse – all of which would appear to be personal rather than owned by a century. Similarly, personal items of armour were located in the third period of use of these blocks: a helmet fitting, a fragment of mail, three chapes, two shield fittings, as well as the remains of a possible iron shield.

The repair of weapons, or the manufacture of small items, is evidenced by composite bows and their bone components. The heads of the weapons listed above might well be in the blocks for wooden shafts to be fitted, which would suggest that some of the rooms were workshops. We have a great deal of evidence that a legionaries' arms were personalised, with much of the armour being custom-made to fit individuals, especially *lorica segmentata*. For the legionary, the cost of replacement of armour and weapons would have been expensive, ensuring that the equipment was well looked after and repaired when necessary, possibly by the soldier himself, especially if he wanted to sell it on after retirement. Demand for arms might have been relatively low in peacetime, since any piece of equipment may have had several owners. There may also have been stores for unused equipment, which might explain a whole suit of legionary's armour being found in a possible storeroom in recent excavations in the Priory Field.

Vegetius comments,

The legion owes its success to its arms and machines, as well as to the number and bravery of its soldiers. In the first place every century has a *ballista* [for firing arrows and stones] mounted on a carriage drawn by mules and served by a mess that is by ten men from the century to which it belongs … The number of these engines in a legion is fifty-five. Besides these are ten *onagri*, one for each cohort; they are drawn ready armed on carriages by oxen; in case of an attack, they defend the works of the camp by throwing stones as the *balistae* [darts].

Further associations with armour are Nos 20/21, which are studs for attaching to leather, Nos 14/15 are decorative buckles, and the other bronze objects are related to leather belt buckles. (*Archaeol Cambrensis*, 1932, Vol. 86, Part 1 p. 84)

These weapons were probably dismantled and stored when not in use, and reconstructed for manoeuvres. In the third phase of the use of the rampart buildings, what looks like the considerable reserve of expendable equipment for a century, or cohort, and called for in times of need included nineteen arrowheads, fifty-six *pilum* heads, three *ballista* bolts, six spearheads, and nineteen slingshot bows. The remains of heavier equipment are evidenced by two catapult bolts heads, caltrops, *ballista* bolts and stone slingshots. There would also need to be a stable for the mules, and at least three sets of rooms with concrete floors and stone-built drains suggest stabling, though it is more likely that these were connected with iron- or bronze-work and that mules would have been stabled outside of the fortress.

While these structures might have been used for the storage of a century's large equipment, workshops for the repair of small personal weapons, or for the last stage of the construction of small arms, it is unlikely that they could provide the necessary amounts of military hardware needed by the legion and the auxiliary forts in its command. This would be the role of the *fabricae*, spread out along the *via quintana*.

4

The *via quintana*

Vegetius tells us that the legion had to be self-supporting, which would imply feeding 5,500 men, as well as clothing them and having a supply of weapons. The implication of this is that a large area of the fortress would have had to be given over to workshops and stores. The problem is that it is not always clear how many workshops there were, or where they were sited, as these types of structures are notoriously hard to identify, especially if an area has only been trenched and not excavated. Hearths might be seen as indicative of a workshop producing iron or bronze, but they are found throughout the fortress as demonstrated by the discovery of a metalworking hearth in a tribune's

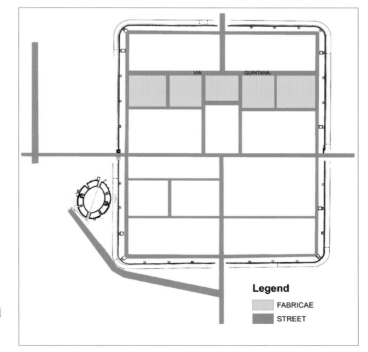

VIA   QUINTANA

Legend

FABRICAE

STREET

This set of *insulae*, stretching across the fortress, has been seen as forming an industrial zone with workshops and storehouses. There is evidence that small-scale industrial activity took place throughout the fortress, and the *insulae* between the granaries and *via praetoria* is possibly a storage area.

house, and the *armamentarium* also had evidence of weapon repair. Items made with leather and wood do not survive, and the scrap metalwork would most probably have been removed from the site when the fortress was abandoned. Tools might help identify the processes, although there is nothing particularly unique about the cutters or needles, which may have been used elsewhere for other tasks. Mark Lewis has observed that any charcoal spreads might easily have come from processes used during the construction of a building or the burning of materials at demolition. Perhaps we should use other criteria to locate these structures, as the evidence from each of the British permanent fortresses is that although the position of the suspected *fabricae* is not uniform, their by-products of soot, smoke and foul smells meant that they needed to be upwind from the high-status *principia* and *praetorium*. At Isca, it is likely that the industrial area forms a band across the fortress from the south-west to the north-west defences. It is separated from the lower-status Prysg Field barracks by the *via quintana*, which is accessed by the *via decumana*, which in turn leads to the *porta decumana* and the legion's land beyond the fortress from which prepared materials such as leather and ingots of iron were probably transported.

At Isca there would appear to have been three types of *fabrica* (storerooms):

i)   A long, rectangular hall, perhaps with a central corridor or veranda, containing rooms of various sizes. The roof acted as a hangar, protecting the activities from the weather and admitting air as well as removing smoke and fumes.

ii) A U-shaped plan, with ranges of rooms or aisled halls that formed sides of a square with three wings and a loading bay.

iii) A large rectangle or square building with numerous rooms, grouped around a courtyard with a large central water tank fed by the fortress main supply.

Clearly, such plans indicate different roles and activities, which, as yet, we can't ascertain.

Vegetius lists the special skills required in the manufacturing area: carpenters, wagon builders, blacksmiths, making the engines of war, repairing arms, vehicles and every sort of artillery. There were also factories for shields, armour and bows, in which arrows and javelins, helmets and every variety of arms were manufactured. Tarruntenus Paternus adds *ballistae*, builders, glaziers, artificers, coppersmiths, shield makers, trumpet makers, and horn makers. An undated papyrus, most likely to be from *Legio II Traiana Fortis*, dating from around the second or third centuries AD, indicates that at least 100 men were employed in the workshops in any one day, producing two sorts of shield, iron plates and catapults. At the northern British fort at Vindolanda, there is documentary evidence that 343 men were in the workshops on a single day, including cobblers and building constructors. More than 1 million nails (measured by weight not tallying) found buried in a pit at Inchtuthil could have been produced by two teams of seven smiths within a year. Outside the fortress, Tarruntenus Paternus details *immunes* such as limeburners, woodcutters, charcoal burners, and to these we should add quarriers, iron makers, tanners, tile makers and brick makers whose products would have been used within the fortress, either directly, such as masons and shinglers, or as part of another processes, such as using the iron ingots to make weapons. All these activities were under the direction of the *optio fabricae*, who was responsible to the *praefectus castrorum*.

THE EXETER *FABRICA*

To get some insight into how manufacturing processes might have been carried out, we need to return to *Legio II Augusta's* previous fortress at Exeter. In 1971, before the construction of the Guildhall Shopping Centre, excavation identified the *fabrica*. Only the eastern corner could be excavated, however, the rest of the structure being destroyed by the commercial development. The front range of what may have been part of a courtyard building with two other aisled ranges seems to have contained workshops and storage lofts with access to the latter being gained from wide, internal loading bays. Next to the north-east entrance and loading bay, was a carpenter's shop furnished with benches. It comprised of an aisled hall 9 metres wide, and containing at least four bays and a room 9 metres by 7.5 metres. The construction of the building was substantial; the post trenches were 90 cm in depth and the posts themselves had been driven into the

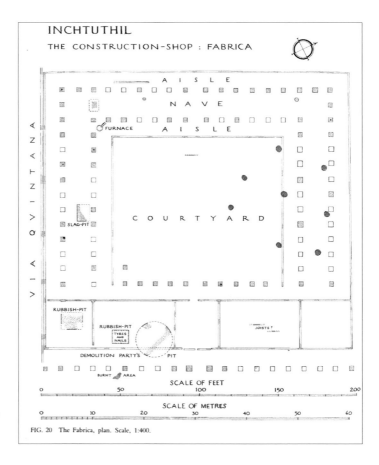

The *fabrica* at Inchtuthil. (*Pitts and St Joseph, p. 104*)

ground a further 40 cm. This suggests that the hall rose to a considerable height, and the nave might have been lit by clerestory windows high up in the walls. Clearly, this space was needed to circulate the heat and smoke from the processes being undertaken.

Inside the aisled hall, there were a large number of shallow troughs, no more than 20 cm in depth and 40–50 cm in width, although the length varied between 50 cm and 4 metres. Their sides were clearly lined with stakes intended to hold plank linings in position. One trough had fine, sandy earth interleaved with much bronze powder and tiny lumps of metal, clearly the result of trapping the waste of lathe turning, engraving or filing bronzes with the intention of re-smelting later. Others were filled with charcoal and bronze scraps, presumably debris from the floor of the hall, deposited after the plank linings had been removed and the troughs went out of use. Bay 1 was partially screened off and its floor covered with charcoal and burnt clay, on top of which was a hearth. In Bay 2 there were wooden troughs, charcoal and debris from bronze making. In Bays 3 and 4 were drains.

The bronze objects from the troughs and layers of debris provide some information about the manufacturing processes in the *fabrica*. There were flat plates with thin shanks, probably rough outs for girdle plates of *lorica segmentata*, and also two complete examples of this type of fitting and three hinge plates, which had been used, and preserved traces of the rivets that secured them to the iron plates of the *lorica*. The majority of metal finds were off-cuts – thin bronze sheets, short lengths, cuts of wire and three bars, the ends of which appeared to have been struck of by a chisel. There were also fragments of two tweezers. Similar damaged or unfinished metal pieces were also found in the *armamentarium* below the Prysg Field rampart at Isca. The importance of recycling scrap in the production of metal goods, such as new armour, might have made the army's need for raw material less than is commonly thought.

An archaeologist specialising in the Roman military, M. C. Bishop, has suggested how *fabricae*, like the Exeter example, might have been used. He has produced two models for the construction of military equipment after the manner described by Tarruntenus Paternus, which emphasises the use of expert *immunes* supervising the less skilled *miles* or slaves undertaking more mundane tasks. This may have been facilitated by the use of standardised patterns, often in the form of templates, that could be traced on the prepared metal or leather surface and produce cut outs. This would allow many individuals to fabricate a small range of items. Bishop offers two examples:

Making a spear

i)   Outside of the fortress, a woodcutter supervises the gathering of wood of the required shape and quality, which is transported by unskilled workers.
ii) A semi-skilled smith forges blades, rivets and butts, the necessary resources – water and fuel – is brought in by unskilled workers.
iii) A skilled carpenter shapes the wood and an expert supervises the assembly of the work by an unskilled worker.
iv) The final product is transported to a store.

Making a suit of *lorica segmentata* armour

i)   One or more smiths prepare iron plates from ingots, with unskilled men at the bellows.

ii) One or more coppersmiths prepare the copper alloy, with unskilled men working the hearth.

iii) Leather workers supervise cutting and stitching the strapping for armour, with unskilled men doing the cutting from patterns and stitching.

iv) Unskilled men use patterns and cut out sheet copper alloy components. Then they assemble and rivet together, with an armourer checking quality.

v)   The final product is taken to store by an unskilled labourer.

Presumably other equipment required by the army – wagon wheels, hubs, oak barrels, gates and winding gear – could have been constructed in the same manner.

LEATHERWORKING

The Roman army required leatherwork on a vast scale, much greater than any other military essentials except foodstuffs and, perhaps, textiles. A list of the equipment needed by the legion, and probably the auxiliary troops (although they would probably make boots for themselves), might include shoes, shield covers, baggage covers and kit bags, saddles and horse trappings, briefcases, letter and tablet envelopes, cushion covers and purses, all of which have been found on military sites. We might also add awnings, smiths' aprons, wine and water skins, flask covers, sheaths and quivers and a considerable amount of straps and fittings for armour. A tent for the eight men of a *contubernium* would take seventy goatskins, while the cover for a shield would take up to two cowhides. If each soldier had three pairs of shoes a year, this would average 18,000 pairs per legion and 1,500 cowhides. Leather is a perishable by-product of stock raising for human consumption, and an excessive slaughter of livestock for making leather alone would have major repercussions for the future supply of food. Even if there were enough hides, tanning is such a slow and long process that there would have been too much advanced planning, which would not sit well with responding to emergencies. Several strategies may have been used together to overcome these difficulties, firstly by ensuring a steady supply of hides, secondly by holding adequate stocks and finally keeping equipment in good repair and also recycling it.

There may have been some hides coming from the legion's own lands, though the tanneries would have been situated well away from any settlement as the use of urine, animal faeces and decaying flesh makes it a particularly noxious procedure. In the auxiliary forts, the men may well have eaten the cows or goats, prepared the cleaned hides, cured them by air drying or salting to stop them rotting and sent them back to the fortress. However, it is difficult not to conclude that the army obtained the huge supplies of leather needed via imports from outside the empire.

Unfortunately, finding archaeological evidence for producing such a wide range of equipment, and trying to interpret it to work out the mechanisms of the supply of the raw materials and the techniques involved, are limited by the very nature of the material itself. Leather only survives in special environmental conditions, which means waterlogged deposits in wells, ditches and river courses (possible pieces of leather aprons and shoes were found in the waterlogged ditch of the fortress in the Prysg Field excavations, indicating that it was no longer a viable defensive element). Such contexts are usually associated with rubbish dumps, and it is very difficult to date them. Leather can also survive in very dry, buried layers, but these are non-existent in Britain. The examples of leather that do survive are almost invariably discarded as worn out or useless, the off-cuts of manufacture, damaged pieces removed and discarded on repair and finally dumped as worthless. It can be extremely difficult to interpret evidence that was consciously thrown away. Similarly, other organic components of the final product, such as glue, wood and stuffing, also disintegrate, making it extremely difficult to put the constituent parts of larger items together as a whole. We have pictorial sources of the manufacture of leather goods and the archaeological finds can help us give these more detail.

At the fortress of Vindonissa, Switzerland, a large and varied collection of tools and off-cuts were retrieved. These indicated the manufacture of footwear and the maintenance of equipment in general, as well as the production of shield covers, tents, straps and other military gear. As Bishop has suggested, it is clear is that there were standard cutting patterns for items such as footwear: openwork, strap-like uppercut with a middle sole sandwiched between the inner and outer soles with the whole thing being nailed together. Even the pattern of nails was identical with no regional variations across the northern frontier, including Britain. It is possible to suggest a process for manufacturing these and other objects; the skilled shoemaker (*immunes*) cut out the pattern for the unskilled soldier who made up the footwear under supervision. It seems that the whole process was under central control with standard patterns available. This is shown with leather off-cuts from a variety of sites demonstrating the arrangements of components, the use of specific seams and reinforcements. This ensured that with careful selection of leather, there would be a minimum wastage in cutting. What also appears to be regulated is the use of cattle hides for large rectangular shields and footwear, but goat hide for tents and round shield covers.

STORES

It is not known where the necessary foodstuffs – cereals, meat, cheese/milk, vegetables, salt – were stored and how they were distributed. Similarly, the stores of exotic varieties of goods only obtainable from abroad – types of pottery, glass, olive oil and wine – have not been identified. Large buildings are the most likely, but *tabernae* along some of the streets could have been used, although it appears that at Isca they did not line the *via quintana* or *via decumana*. Products being obtained through army contract or sold privately in the *canabae* are both possible options, but whatever the source, staple foods would have been deducted from the individual legionary's pay, or perhaps shared with other members of his *contubernium*.

5

The *via principalis*

One of the centurions from the Prysg Field cohorts walking along the *via sagularis* to the *principia* for a meeting with the *primus pilus*, one of the tribunes, or an assembly with the legate, would turn into the *via principalis* and the rows of colonnades on either side would draw his eyes to the *groma,* the ceremonial entrance to the headquarters building.

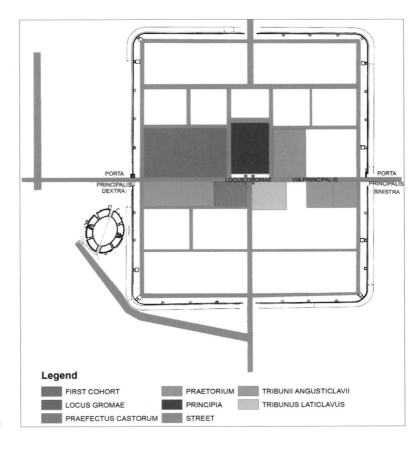

The *via principalis.* (Caro McIntosh)

Legend

FIRST COHORT	PRAETORIUM	TRIBUNII ANGUSTICLAVII
LOCUS GROMAE	PRINCIPIA	TRIBUNUS LATICLAVUS
PRAEFECTUS CASTORUM	STREET	

COLONNADES AND *TABERNAE*

The colonnades along each side of the *via principalis* were probably about 4 metres wide, and kept people dry as well as having an aesthetic effect. The colonnades acted as porticos for the *tabernae,* spaces with approximately the same dimensions – 8 metres deep and around 4.5 metres wide – sometimes divided into two and reminiscent of shops in towns. They are enigmatic buildings, as two examples at Caerleon demonstrate. Excavations at Broadway House in 1994 on the left side of the *via principalis* had a *taberna* in both in the timber and stone phases and, from the mass of *amphorae* shards and animal bones, it seems the area was associated with the sale of foodstuffs, perhaps even a tavern. However, in 1987, an excavation beyond and on the opposite side from the *principia* indicated that there was another row of *tabernae* behind a portico, also 4 metres wide. These buildings yielded evidence of iron, bronze, and lead working, as well as fragments of armour, some unfinished. At Inchtuthil there is evidence of the use of *tabernae* as storerooms and workshops, and elsewhere glassware and pottery have been recovered. They may also have held weapons, tools, tents and leatherwork. It is most likely that the *tabernae* would have had doors that could be closed for security, or to make a more pleasing backdrop for official, formal processions to the parade ground. Not only did the colonnades give an architecturally balanced and complete appearance to the street, but they also hid the quarters of the higher echelons of the legion behind them, giving those of high status some privacy.

THE FIRST COHORT

Hyginus is the only classical source for the first cohort, and tells us that it was formed of double centuries (160 men), occuping twice the area of the other cohorts, and its double number of barracks were to be found on the *via principalis,* next to the *principia*. The centurions of the cohort were senior to the others in the legion, and so were collectively known as the *primi ordines*. The arrangement of their ranks was also different from II–X, and the arrangement was:

> *Hastatus posterior*
> *Princeps posterior*
> *Hastatus*
> *Princeps*
> *Primus pilus* and senior centurion of the legion (possibly with two centuries)

The *primus pilus* (chief spear) served for one year, and was likely to be promoted to the post of camp prefect in another legion. The other centurions of this elite unit were probably promoted through the cohorts, and the members of the centuries picked from others across the fortress. The *aquilifer* (legionary standard bearer) and *imaginifer* (bearer of the image of the emperor) were part of the first cohort.

The difference in ranking is probably reflected in the size and facilities of the centurions' quarters, which appear to have been detached from the rows of *contubernia*. Excavations, and later geophysical prospecting at Caerleon, suggested that the structures had courtyards that may have contained gardens, and were twice the size of those of the Prysg Field centurions' corridor accommodation. The houses were divided into small living rooms and equipped with elaborate systems of drains, were mostly floored with concrete and the walls decorated with painted plaster. The *primus pilus* had the largest house next to the *principia*, as he was in charge of the standards and was the direct link between the centurions and the legate.

THE *PRINCIPIA*

If the *via praetoria* was about the legion's image of itself, then the *via principalis* was about the status of the hierarchy who controlled the fortress and its administration. It could be called the 'street of purple' from the bands that decorated the togas of the tribunes and the legate. It was the only part of the fortress where women, the wives of the senior officers and their servants were allowed. Where the two streets came together, both roles were combined

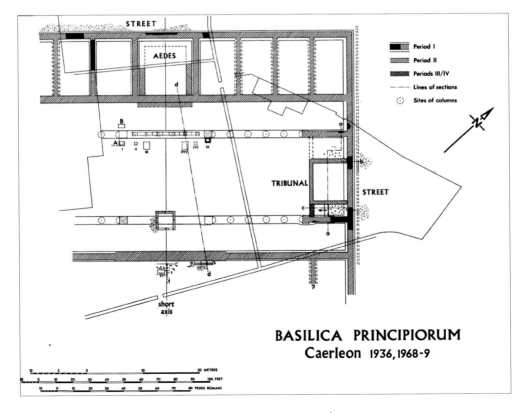

The stone *principia* at Isca. (*Archaeol Cambrensis, 1970, Vol. 119, p. 14*)

in the same building, the *principia*, in the plural in Latin, which reflects the complex function of the building. The origin of the *principia* comes from the Republican army's temporary camps where an open space (the *opera*) was sited next to the commander's tent (the *praesidium)* in order for the assembly of the troops. Eventually, the two functions were formalised in permanent fortresses by the *principia* (headquarters) and the *praetorium* (the legate's accommodation), but perhaps the divide between their functions is less than is usually accepted, and in many ways they still formed a complete entity.

The stone *principia* at Isca was approximately 93–95 metres long and 66.2 metres wide, covering an area of around 0.62 hectares, about 3 per cent of the fortress. A vaulted passage through the monumental entrance of the *groma* led to a large, flagged courtyard, continuing the choreographed view from the *via praetoria* into the *aedes* (Chapel of the Standards). The courtyard was surrounded by colonnades on three sides, covering offices. The fourth side was closed by the façade of the *basilica principiorum*, the form of which is not known at Isca, but at Novae was a line of arcades with the central entrance on the axis of the complex. The *basilica* comprised a large hall, 66.2 metres by 29.0 metres, with four free-standing columns 10.2 metres, high which divided the space into a wide nave with two aisles. These columns held up a roof 20 metres above the floor – the *basilica* at Eboracum was 25.7 metres high with the remains of the *principia* having been excavated below the cathedral.

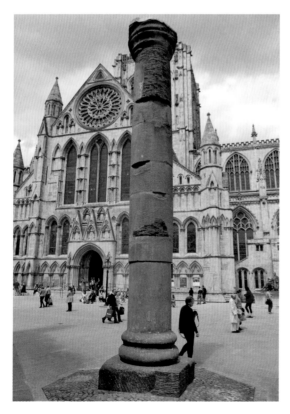

The reconstructed column from the *principia basilica* discovered under York Cathedral. It is 7 metres high and 1 metres in diameter. (*Don Henson*)

Once inside the *basilica* at Isca, there are two massive arches opposite each other on the long axis of the hall, springing from huge piers 14 metres apart. These interrupt the row of columns and thereby manipulate the eye to the *aedes*. Underneath the first great arch, and on this central axis, was the square outline of large, white sandstone blocks on which were set rails with a gate on one side, and which was interpreted as an altar enclosure for a statue. Although the excavator suggested that coins and a few fragments of animal bone found within the square of stone have no significance, animal bones have been found in several *principia*, including Inchtuthil and Novae, where the burnt bones of sheep or goats were interpreted as the remains of ritual banquets, or sacrificial pits filled with an offering before construction began.

Along the opposing short sides of the basilica were two tribunals, or podiums, where small assemblies were held, presumably for centurions for the issue of orders, the passing of the watchword or the hearings of military courts. There may have been a doorway from the side of the north-eastern one into the legate's residence, as the space beyond the right *tribunal* was wider than the other and was filled with rubbish, which may have accumulated through a doorway into a passage. Unfortunately, the excavator's trench ended just before this possible entrance. An opposing entrance may have been for the *primus pilus* of the prestigious first cohort, which was responsible for the standards and setting the watchword for the day. At Novae, beneath the tribunal, was a pit for altars that were no longer in use. At Isca, the walls were trenched, but no such objects were found.

Under the further of the two large, high arches was a screen with six columns not more than 3 metres high, defining the sanctity of the space behind it, but not blocking the view into the *aedes*. In front of the screen were the possible bases of six statues, at least one of which was superhuman sized in bronze armour, probably an emperor. The cult of the emperor, *Numinibus Augustorum*, was the official religion of the Roman army and dominated its sacred life. Many emperors were made into gods after death, and their images were also displayed in the *principia*. At Novae, the statues included at least five emperors and deities, including Jupiter, Victory and Bonus Eventus (good luck), and there is evidence of other statues from inscriptions on altar bases. Behind the screen and statues was the *aedes*, the religious focus of the legion. This raised room had a back wall dedicated to statues and emblems, the most important of which was the eagle. An object of worship, it was made of bronze and plated with gold, with outspread wings, grasping a thunderbolt in its talons and was mounted on a wooden pole. There would also have been the *genius* (spirit of the legion). At Isca, two of the walls had low benches (*suggestus*), probably for statues dedicated to deified emperors and with the standards themselves probably at the rear. At Novae, at the back of the *aedes*, were seven stone blocks with holes, most probably for the wooden podium, where the standards were put. On the back wall of the chapel was a stone tablet with an inscription detailing the renovation of the *aedes*. Something similar may have been in the same position at Isca as a fragment of such a plaque was found in a robber's trench 8 metres from the *aedes*. The translation, restored by the excavator George Boon and others, might have read:

To Jupiter Best and Greatest and to the Guiding Spirit of the Emperors Antonine and Commodus, Titus Esuvius —, Legate of the Emperors, restored the Temple of the Standards —, Chief Centurion, gave and dedicated it.

It seems that over time the *principia* was becoming more of a temple to the legion's religious cult. Either side of the *aedes* were probably the legion's administrative centre, the *tabularium legionis*, staffed by the clerks (*librarii*), under the supervision of the adjutant (the *cornicularis*) and his orderly (the *beneficiarius*).

ROME IN THE FORTRESS

The position and arrangement of the high-status buildings along the *via principalis* echoes the political life of Rome. At the top of the social ranking were the senatorial and equestrian classes, differing largely on property-based criteria. The senatorial order was the richest, with individual wealth coming mainly from property and large agricultural estates. They wore the *laticlavius* (the broad reddish-purple stripe) on their togas. The equestrians were largely engaged in commerce, and this influential business class was identified by narrow stripe (*angusticlavius*) of the same colour. The division between the two classes was a boundary that could be crossed if equestrian families could assemble sufficient resources for their descendants to qualify for entry into the Senate. Both classes had career paths for their male offspring, which included the army, in order to give them the experience they would need for their public lives in their social rank.

The senatorial career path (*cursus honorum*) involved service as an *augur* (priest), at least a year as a tribune with a legion, and then a civil service post before becoming a legionary legate. The *legatus Augusti pro praetore* was the personal appointment of the emperor, but with only thirty legions at any one time, this would have required a lot of patronage. The legate was usually in his forties, serving in the legion for the standard period of command of three years. He might then go on to to be the governor of a large province. We know part of the career of Tiberius Claudius Paulinus, who was the legate of *Legio II Augusta* in the reign of the Emperor Caracalla (AD 211–17). Subsequently, he held office in two Gallic provinces, before returning as governor of Britannia Inferior, the most northerly of the two parts of the island as divided up by the Emperor Septimius Severus (AD 198–211). The second-in-command was also from the senatorial class (the *tribunus laticlavius*), probably about twenty years old and serving for a year.

The equestrian path was known as the *tres militiae* (three military posts), a career progression in the Roman Imperial Army. It developed as an alternative to the *cursus honorum* of the senatorial order, to enable the social mobility of equestrians and identify those with the aptitude for administration. The three posts, each typically held over a period of two to four years, were as a prefect of a *cohorts quingenaria*, one of approximately 150 auxiliary infantry units of 500 men, then the military tribune of a legion, and finally prefect of an auxiliary cavalry *ala* (wing), which

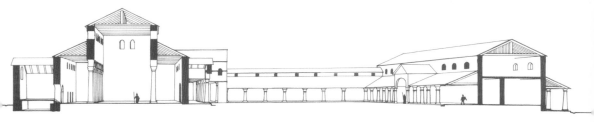

A reconstruction of the Chester *principia*. From the right can be seen the entrance and offices, across the courtyard is the *basilica*, behind which is the *aedes*.(*Copyright of Cheshire West and Chester Council*)

protected the flanks of a legion. However, because of the shortage of legionary tribune posts, an alternative for the second *militia* was as auxiliary tribune with a *cohors milliaria*, one of thirty regiments of 1,000 men each. In some exceptional cases, a man might receive a fourth promotion as prefect of 1,000 man *ala,* though fewer than ten such *alae* existed. An equestrian would be in his mid-thirties or older at the conclusion of his *tres militiae,* and would either have a career in politics or business. Some men instead moved on to posts in imperial administration, especially as *procurator*, the chief finance officer of a province.

There were five equestrian tribunes with the rank of *tribunus angusticlavius,* the military tribune of the narrow stripe. The status of each of these roles can be detected in some fortresses by the position and size of the accommodation. The roles of specific tribunes are not distinguished by available literary or epigraphic sources, but the *Digest* indicates administrative and judicial responsibilities rather than those of direct command:

> Keep the soldiers in camp, to lead them on exercise, to hold the keys to the gates, to make occasional inspections of the pickets, to attend the distribution of corn, to inspect the corn, restrain dishonesty in those who measure it, to punish offences in accordance with the measurement of their authority, to attend frequently in the *principia*, listen to disputes of their fellow soldiers and to inspect the sick.

Clearly, there would have been a lack of military experience with these young tribunes, and an important career soldier, the *praefectus castrorum* (camp prefect), a fifty- to sixty-year-old man who was of equestrian status and third in command, had the responsibility for the daily running of the fortress. He was accountable also for the surveying and construction of the fortress and the forts in its command, as well as being in charge of the workmen. The importance of this rank was reflected in position of his house, usually facing the *principia* and opposite the *tribunus laticlavius* across the *via praetoria*. This post was filled by a previous *primus pilum*, but probably promoted from another legion.

THE *PRAETORIUM*

The legates' accommodation was behind, alongside (as at Isca), or even across the road from the *principia,* which emphasises the close relationships that may have been further enhanced by the *praetorium* being accessible from the cross hall of the *principia*. This suggests that the headquarters could in some cases have acted as the main entrance to the legate's quarters. The legate was from the senatorial class and therefore would have expected a high standard of luxury in his accommodation. Vitruvius, in his *On Architecture,* suggests,

> For distinguished men who are obliged to fulfil their duty to their fellow citizens by holding public office and magistracies, we must build lofty vestibules in regal style, and spacious *atria* and peristyles … also libraries and *basilica* comparable with the magnificence of public buildings, because very often public meetings and private trials and judgments take place within their houses.

What he doesn't mention is the need for the public to see through the house, so that the wealth and power of the owner could be recognised, or that it needed to be seen from a significant part of Rome, such as the *Forum Romana*. The *praetorium* was in a significant place and the legate was known to be powerful, therefore the need for external display was superfluous; it was privacy that was needed. However, because he was from a wealthy background and established social position, he would have lived in a style not far removed from the customary one. There needed to be substantial apartments for entertaining, and private rooms for his wife and family, as well as accommodation for a large staff (probably brought with him like his furniture and belongings) – although this might not be housed in the *praetorium,* but nearby. The difference between senatorial accommodation in Rome and that of the legate in a fortress was that of many of the features described by Vitruvius – the fulfilling of duties, the library, the judgments – would have taken place within the *principia,* which may be seen as the replacing the traditional entrance rooms of the *vestibulum* and *atrium* of the *praetorium*.

Identifying the *praetorium* in British fortresses is difficult, however, the key attributes of an elite house should be its size, position and especially how the structure was adapted from Mediterranean styles to the British climate. Again we are reliant on the fortress of Isca for the details, but even here two buildings have been suggested as the legate's residence. Behind the *principia* was a structure with a large pool, and this was compared to the one at the known *praetorium* at the German fortress of Vetera. However, there the resemblance ends·there and, as Mark Lewis has pointed out, the site of the legate's accommodation is almost certainly next to *principia* on the *via principalis*, opposite the entrance to the baths *basilica*. The site satisfies the criteria of space covered (probably 50 metres by 80 metres), and although it is smaller than the *principia,* it is almost twice as large as the senior tribune's house across the street. That said, the *praetorium* at Isca was many times smaller than a senatorial property in Rome.

Evidence from an excavation, which trenched the area in advance of the churchyard extension in 1908, suggests that the most luxurious part of the building was in the

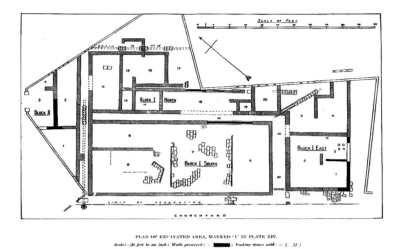

PLAN OF EXCAVATED AREA, MARKED 'I' IN PLATE XIV.

Scale:—30 feet to an inch : Walls preserved : — ▉ *: Foundation-stones only: — I, II |*

The *praetorium* at Isca, excavated in 1908. The axis of the building is on a left to right line through the open area of Block One South, which had a corridor around it. The other half of the building is below a graveyard. (*Bosanquet & King, 1909*)

south where the sunshine would have warmed a peristyle courtyard 15.24 metres wide and 30.48 metres long. It was open to the sky and surrounded by a low wall with columns holding up a roof under which ran a colonnade. The courtyard appears to have been divided into three elements. One at the south end was recorded as being of 'natural clay', perhaps suggesting a garden. In a central position was a sandstone-slabbed area and finally, close to a herringbone brick-floored area, was a 'curious polygonal projection' that may have been a garden feature or statue base. However, an area of brick with the same design and location was glimpsed in the suspected *praetorium* at Chester, and if this was a standard characteristic, then it could be related to a feature known as an *auguratorium*. Hyginus describes it as being on the main street at the right-hand side of the *praetorium*: 'the general might get the augury properly thereupon a tribunal is placed to the left so that on receiving the augury he may ascend and greet the troops with favourable omens.' In a permanent legionary fortress, this ritual might have been relocated to the *praetorium* and the legate delivered the omens on the tribunal in the cross hall of the *principia*.

The corridor around this courtyard had a number of rooms leading off of it, with remnants of the hypocaust underfloor heating system, suggesting rooms with a domestic purpose, such as bedrooms. The only example of an upstanding wall had two layers of plaster on its interior, with purple stripes on a white background. No doubt successive legates would have decorated these quarters to their own tastes, considering the three-year length of their service. This end of the building backed on to the colonnade alongside the *via principalis*, and a blank wall would have ensured privacy, with light coming into the building through the courtyards. There is no evidence that the external walls of the building had any form of decorative display. The excavation of 1908 was hampered by the pre-existing cemetery, but earlier gravedigging retrieved a patterned mosaic of white-and-blue limestone *tesserae*, the individual stones making up pattern, which was of the quality required for a luxurious *triclinium* (a dining room). A similar white mosaic was recognised in the presumed *praetorium* at Chester. On the same side of the house, the discovery of hollow ceramic *voissours* (hollow clay bricks) suggest a vault over a bath building as, being light, they would be easy to support, and also would have stopped condensation dripping onto the heads of the bathers. The unexcavated northern area would have been in the shadow of the cross hall of the *principia*, and must have had an *atrium* to let in light. Surrounding this was possibly the service area (including the *culina*, the kitchen) which, although having access to the *triclinium*, would have been entered by the servants from an outside side street.

TRIBUNES' HOUSES

The tribunes' houses were on the right-hand side of the *via principalis* in the *scamnum tribunorum*. Although their design was similar, the space allowed varied by rank, with the senatorial tribune and the camp prefect (the second- and third-in-command) being accommodated in larger houses than the others, and the former being slightly larger than the latter. Those that were excavated at Inchtuthil appeared to have a portion of the house set out as domestic accommodation with a vestibule, dining room, baths, and kitchen, while the remainder of the building was used as offices. In the centre was a peristyle courtyard, imitating houses in Italy. It is clear that the tribunes' own spaces were designed to individual tastes and requirements, much as the centurions' accommodation in the barrack blocks. The houses appear to have been entered from a lane at the rear, thereby giving privacy, as no barrack blocks impinged on this back street. Part of a tribune's house has been excavated on the museum site at Caerleon, and a fragmentary wooden tablet records a work party sent out to collect timber for some construction project, which suggests that it was the house of the camp prefect. Evidence of metalworking was retrieved in the form of crucibles for the smelting of silver and one for gold – clearly a matter for security, and later a large room was given over to a metallurgical function for the production of bronze and iron. The *tabernae* in front of the tribunes' houses may well been offices for large number of *librarii* (clerks), *cornicularius* (adjutant) and *beneficiaries* (orderly) of the *officium tribune laticlavi*, the *officium praefecti castrorum* and the other lesser ranked tribunes.

When the baths *basilica* were excavated, it became clear that a structure had been demolished on the *scamnum tribunorum* to make way for it. It has been suggested by the excavator that this might have been the home of the commander of the *Ala I Thrace*, closely identified with the legion, and whose equestrian rank would have been appropriate for such a prestigious position. There is no known area for the stabling of horses or accommodation for cavalry in the fortress, or the mules and oxen for each century, and they may have been kept outside the walls. However, each of the senior officers would have needed several mounts, and these may possibly have been kept in nearby *tabernae*.

6

Community Spaces

BATHS: *THERMAE* AND *BASILICA THERMARUM*

The legionary would have experienced training in rivers and streams, so he had confronted nature at its wildest and most unpredictable. However, the baths complex demonstrated that nature could be controlled through Roman ambition and technology, and that huge quantities of water could be channelled, contained and heated, as well as moulded, sculpted and used as a vehicle for decoration. The *thermae* at Isca had its *frigidarium* (the cold room) comprehensively excavated, but the remainder of the complex continues to be covered by medieval and modern structures. The *tepidarium* (warm room) and *caldarium* (hot room)

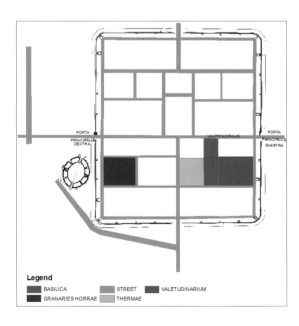

The *thermae* and *basilica thermarum*.
(*Caro McIntosh*)

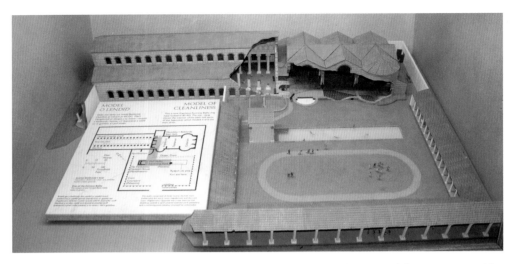

A model of the *basilica* and *thermae* demonstrates the monumental nature of these structures. The plan to the left indicates in green the area of excavation at Isca with the remaining detail being from other sources such as Exeter and Chester. The aisled *basilica* is on the left and the *thermae* is arranged from the left: *frigidarium*, *tepidarium* and *caldarium*, the latter two having underfloor heating from furnaces. In front is the *palaestra* with the *natatio*. (*Author's collection*)

of the *thermae* at Isca/Exeter have been uncovered, which, although smaller, were built by the same legion ten to fifteen years earlier. The baths complex at Deva, although laid out differently, has similar dimensions to those at Isca. Using the evidence from all three fortresses, we can get a grasp of the monumentality and workings of these extraordinary structures.

At Isca, the main entrance to this grouping of structures was on the *via praetoria*, although there was another on a minor street at the opposite side with access from the hospital building. Because the *thermae* building was set back around 60 metres from the street, it unlikely that it would be seen over the colonnades that lined the *via praetoria*, so when the legionary entered the complex he would be confronted by the cathedral-sized monumental structure of the *thermae* and, across a courtyard, the *palaestra*. Because this enclosed space had a portico surrounding it on the other three sides, and perhaps only the roof of the *principia* was visible from within it, the effect of would have been that of an island of Roman luxury within the austerity of the fortress, and the drab confines of the rudimentary and often sordid *contubernia*.

THE *PALAESTRA*

The courtyard was an enclosed space of 3,224 square metres (62 metres by 52 metres) with an open-air swimming pool (*natatio*). The *palaestra* could be used both for outdoor games, especially as it was facing south to catch the sun, and for training, as its sand and gravel floors provided a more secure surface for hobnailed boots than the sandstone flagstones that were used in the *thermae*. Perhaps a more prosaic use of the *palaestra* has been conjectured at Exeter, where a circular feature was interpreted as a cock fighting pit, and indeed spurs from cockerels were found in the fortress drain at Isca.

The *natatio* had a richly decorated, shrine-like fountain house at its north-west end. This *nymphaeum* may have had a group of statues centred around Venus rising from the sea, of which only a stone dolphin acting as the fountain head remains. The goddess, besides having the attributes of love, beauty, sex and fertility, was also seen as the mother of the Roman people, and Julius Caesar and the other emperors of the Julio-Claudian dynasty claimed her as an ancestor. Augustus was part of that line and so Venus would have been the protectress of *Legio II Augusta*.

THERMAE

The *thermae* provided a cumulative series of sensory encounters induced by the fire-fuelled rising and falling temperatures of air and water. The climate in each of the rooms might have been ritually experienced several times, with the bather moving from the *frigidarium* to the *tepidarium* and finally the *caldarium*, and then retracing his (or her) steps through the sequence to play outdoor games in the *palaestra*.

Each of the three main components of the *thermae* was in line along the axis of the building. Each room was symmetrical and the same size, with walls in between to keep temperature zones separate. At Isca, the foundations were of rubble and mortar, 1 metre deep, with the walls on top constructed of rubble cast with a neat facing of small, dressed stones. The walls at the ends of the building were particularly massive, and acted as buttresses to take the outward thrust of weight of the concrete vaults forming the roof. At Deva, it has been estimated that the concrete barrel ceilings soared to 16.1 metres above the floor. These vaults would have had to be built together to ensure that the pressures were equal, and therefore they supported each other. A wooden formwork would have been used to support the arches during construction and, when the vaults were complete, the scaffolding would have been used by the artists as they worked down the walls painting the ceilings and columns.

At Isca, a changing room (*apodyterium*) led to the *frigidarium,* which housed two cold plunge baths (*piscinae*), as well as *labrae* (circular wash basins) in alcoves either side. In the centre of the stone floor was a drain to take surplus and waste water away. This room was probably where most of the dirt from the body was cleansed using *strigil* (a curved metal implement that fitted the contours of the body). Although the coldest room, and probably the dampest, the walls were covered in paintings, as was the ceiling of the vaults, and carved stone was used for decoration.

Enough of the *tepidarium* at Deva was recorded, during its destruction for the building of a shopping development, to show that this room was completely covered with fine mosaics and decorated with painting and carved stone. As at Isca/Exeter, a hypocaust system existed below the thick floors, the gases now cooled with the heat having been dissipated below the hot room. Any of the hot, dry air escaping through the door from the *caldarium* would have also helped to keep the *tepidarium* warm, and there may have been an ancillary furnace to maintain the temperature. Presumably, it was in the comfortable warmth of the *tepidarium* that most of the social life of the baths took place, partaking in food and gentle social activities, worlds away from the usual experience of the soldier.

The *caldarium* with its hot, dry, sauna-like heat would have been used to encourage perspiration, and it is likely that wooden-soled clogs were worn in here. In this space, two recesses reflected those of the *frigidarium* and probably had *labrae*, but this time provided with hot water, and possibly a statue in between them. Either side of the room were two plunge baths (*alveii*), again filled with hot water. The use of oil to further cleanse the body using a *strigil* would have been very effective. Hot gases were produced directly from the fire of the main furnaces, and ducted through flues that ran under the *caldarium* floor, and then through arches into the *tepidarium*. This was then piped up the walls and across the ceiling vaults through a continuous lining (*tubulatio*), made up of hollow box-shaped bricks (*tubuli*). This resulted in all the surfaces – floor, walls and ceilings – being heated, thereby stopping condensation forming on the floor and onto the users of the other rooms. The gases were then vented through the vaults, either at eaves level or through chimneys, and in turn this produced a draught, which ensured that the hot fumes were spread evenly throughout the hypocaust. The hot water that fed the *alveii* and *labrae* was produced from a boiler, which sat on iron supports above the fire, a stopcock being used to control the temperature of the water, which was abstracted from the main fortress system into a reservoir (*castellum aquae*).

In order to quantify the amounts of water used in the Isca baths in each twenty-four hour cycle, it is to the Deva *thermae* with its the similar dimensions and the work of David Mason, that we must turn.

FACILITY QUANTITY	LITRES	GALLONS
Basilica		
i) drinking fountain	12,000	2,639
Palaestra		
i) *natatio* (refilled daily)	302,400	66,518
ii) fountain	55,000	12,098
Frigidarium		
i) *piscinae*	55,000	12,098
ii) *labra*	55,000	12,098
iii) latrine and drinking fountain	12,000	2,639
Tepidarium		
i) drinking fountain	12,000	2,639
Caldarium		
i) *alveii*	66,264	14576
ii) *labra*	55,000	12,098
Total twenty-four-hour cycle	624,664	137,406

The water was probably tapped off the main system at night when demand elsewhere in the fortress was at its lowest, most likely through wooden pipes joined by iron clips. The amount of fuel used is impossible to estimate at present; the experimental archaeology carried out on this sort of process has only been undertaken on smaller, villa-type buildings. However, it must have been several tonnes a day, which has major implications for the timber management of the area and the amount of labour needed. With the proximity of the coal measures to Isca, it may well be that when the furnaces are excavated it is this form of fuel that will be found.

A LEISURELY EXPERIENCE

Without doubt, the baths were constructed for hygienic motives, to keep the legionary clean and fit; however, just as important, they were also for rest, recuperation and leisure. The Roman writer Seneca, in a letter to his friend Licilius, supplies us with a unique account of the activities in a small public bathhouse. It is a well-known piece, and it does not need an apology to use it again here as it is helps in understanding the importance of bathing as an everyday attribute of being Roman.

> I have lodgings right over a bathing establishment. So picture to yourself the assortment of sounds, which are strong enough to make me hate my very powers of hearing! When your strenuous gentleman, for example, is exercising himself by flourishing leaden weights; when he is working hard, or else pretends to be working hard, I can hear him grunt; and whenever he releases his imprisoned breath, I can hear him panting in wheezy and high-pitched tones. Or perhaps I notice some lazy fellow, content with a cheap rubdown, and hear the crack of the pummelling hand on his shoulder, varying in sound according as the hand is laid on flat or hollow. Then, perhaps, a professional comes along, shouting out the score; that is the finishing touch. Add to this the arresting

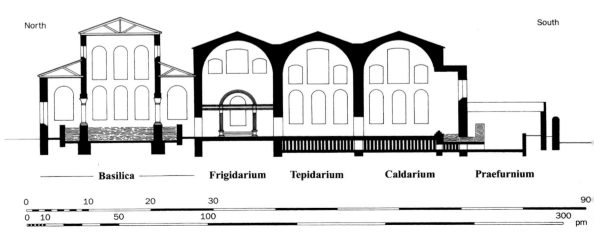

A cross section of the Deva baths. (*Copyright of Cheshire West and Chester Council*)

of an occasional roisterer or pickpocket, the racket of the man who always likes to hear his own voice in the bathroom, or the enthusiast who plunges into the swimming-tank with unconscionable noise and splashing. Besides all those whose voices, if nothing else, are good, imagine the hair-plucker with his penetrating, shrill voice, – for purposes of advertisement – continually giving it vent and never holding his tongue except when he is plucking the armpits and making his victim yell instead. Then the cake seller with his varied cries, the sausage man, the confectioner, and all the vendors of food hawking their wares, each with his own distinctive intonation.

(*Ep. Ad Lucilium*, 96)
Translation by G. C. Boon

How might this compare to what happened in a legionary bathing establishment? In terms of the use of the facilities there would be likely to have been few differences. The presence of so many coins found in the *frigidarium* drain suggests payment for masseurs, food, hair removal and bath oils. Certainly the space under the vaults would have amplified the sounds as such an open, high and solid structure would have caused considerable echoes and, presumably, legionaries had bigger lungs, needed more intensive physical exercise, and had more testosterone to work off. There would no doubt have been different sounds and volumes from each of the rooms. However, a unique deposit in the drains at Isca can tell us a great deal about the more tangible aspects of the bathing experience. After the baths had shut, the *piscina* and *alveii* needed to be drained and refilled and the floors cleaned, most likely by slaves working by lamplight. Any water that was on the floors of the hot rooms was swept down into the *frigidarium* drain, and because of the low level of light, any item accidently dropped in one of the baths, or intentionally discarded, would have been swept into the main drain. The *frigidarium* drain was to become an important source of finds and information, and acted as a 'culture trap'.

As one might expect in a bath building, many fragments were found from the easily smashed with slippery fingers, such as oil containers, which varied in quality from ornate, engraved, colourless glass flasks, the metal chains of which were attached to the stoppers sealing these more expensive bottles, to simple pottery dishes. *Strigili*, formed of a curved blade and handle for scraping off the dirt and sweat after oiling, were also common, and were an indicator of wealth and discernment, from a seashell to a copper scraper inlaid with silver and gold, telling the story of the Labours of Herakles – a very rare find and perhaps indicting that its owner had been of high status. Hygiene was important to the Romans and the recovery of nail cleaners, needles, spatulas and ear spoons should not be a surprise. Military equipment in the form of hinges and rivets, scales, pendants and harness fittings from *lorica segmentata* were retrieved and, since it would be unusual for legionaries dressed in armour to come into the baths, we have to consider whether they were from activities in the *basilica* or *palaestra*, or even small-scale repair or manufacture of jewellery by an entrepreneurial metal worker.

Identifying personal items and allocating them to a gender is fraught with difficulty (and sometimes dangerous) and therefore one has to be cautious, trying not to make inferences from their modern-day usage. Having said that, beads, brooches and dress hooks may have been used by both sexes. Personal female objects may be suggested as ornaments comprising of beads, an earring, and pendants (one of gold). The *intaglios* (etched semi-precious stones from finger rings) are one of the wonders of any of the Caerleon excavations. Although a few stones were found still set in their rings, most of the eighty-eight *intaglios* recovered had been detached. The differential expansion and contraction of their mounts as the temperatures changed in each of the rooms, must have been a major factor in the separation of stone and metal, aided by the hot and damp conditions dissolving the natural adhesive. Some of these gemstones were so small, one being just 4.4 mm across, that they would not have been noticed (or worth picking up by others as they were such personal items). As a group, the *intaglios* indicate changes of fashion over the time when the baths were in use. From AD 75–110 transparent rings (amethyst, quartz, etc.) were common, but in the second phase (AD 160–230), opaque rings seem to have been in vogue. The themes of the etchings reflect other samples found in military contexts, and highlight the everyday hopes and fears of the legionaries. An analysis of the themes suggested a number of categories. Fortuna, Ceres or Bonus Eventus (good events or good luck) were the subject of nineteen gemstones, and thirteen stones had symbols representing deities of fortune and prosperity – ravens and parrots, poppy heads, corn ears and baskets. Bacchanalian scenes with satyrs and the deities Bacchus and Dionysos were found on ten stones, and Cupid, the god of love, on just three. Military deities and symbols were the largest group with twenty-four found, featuring the twins Castor and Pollux, known also as the Dioscuri from the star constellation of Gemini, who were associated with sailors and horsemanship, as well as a charioteer and a horseman. Mars, heroes, a legionary eagle, and winged Victory represented the infantry, and Jupiter, the chief god, was also a favourite. Perhaps the most exciting prospect is that the excavator calculated that there was no fall off of the incidence of finds with the distance in the drain, which suggests that more treasures of this type remain to be discovered.

How did such precious (in terms of personal sentiment and unlikely to have been of monetary value) arrive in the *frigidarium* drain? Losing the gemstones must have been a common experience, but why did soldiers still wear their rings to the baths? The individual male might have felt exposed to the evil of mischievous spirits by their nakedness, which were warded off by the power of the symbol on the ring. Perhaps the most prosaic explanations are that the owner didn't want to leave them in the changing rooms to be stolen, or was it that the ring just wouldn't come off the finger?

In many public bathhouses, writing equipment such an iron or bronze *styli* have been found, indicting a scribe was paid to write letters. However, there was no such evidence of this from the drain, indicating that the men were literate, the necessary accounts were done by the centuries' *librarius*, or that being so far away from home, contact with families was lost. Items for recreation, such as counters, seal boxes, weights, dice and coins, were common. Weaving equipment and a spindle whorl demonstrate time being used profitably, but not necessarily by women. Although it is impossible to suggest from physical evidence that medical treatments were offered, several teeth demonstrate indicate the presence of a dentist.

Because the sediments in the drain had dried out over time, evidence of soft tissue such as Seneca's cakes, sausages and confectionery no longer exists. However, the range of other foods is wide, all having the common factor of being in small, easily held portions.

Animal bones from chicken joints, mutton chops, pig ribs and trotters were all present, and if civilian traders were not allowed in the baths, then this material might have been bought in one of the *tabernae* along the *via principia*. Similarly, there was an assortment of wildfowl, probably the result of hunting along the local rivers or in the marshes, as well as shellfish. Long-distance trade was indicated by an olive stone. These eating habits are reflected in the types of pottery containers found – small bowls and dishes as well as drinking vessels were common, and there was a marked absence of storage or cooking pots. No doubt the legionary would have expected to buy something more palatable than the *acetum* (sour wine), or *posca* (wine and vinegar with water and flavoured with herbs). He may also have expected more than the *cervesa* (ale or beer). The finds give some information about the activities that were taking place, but who was enjoying them? In the second phase of the use of the baths, there is evidence of women and children, and the unofficial wives of the legionaries living in the *canabae* outside of the fortress, and this might be reflected in the 'family-size' joints of meat that occur at that time. What we do not know is how many times a week any particular group of legionaries were allowed to use the baths, whether this was by century or cohort, and whether the upper ranks, which probably had their own small baths, used the fortress *thermae*, although the Herakles *strigil* suggests wealth. Mixed bathing was unlikely, and time must have been allotted to women and any other male civilian.

THE *BASILICA THERMARUM*

The *basilica thermarum* was the exercise hall, rather like today's gyms for personal fitness, but also for training if the weather was poor outside. At Isca it could be entered through the colonnades of the *via principia* or from the inside of the baths. The *basilicae* at both Deva and Isca were approximately the same size – around 64 metres long and around 25 metres wide, with an area of around 1,530 square metres. Both had seventeen columns, with centres approximately 3.5 metres apart, supporting arcading to carry the walls holding up timber frames of the sloping roofs. At Chester, it has been suggested that the height of the nave ceiling was about 17 metres above the floor, close to the estimated height of the baths. At both the floors were of sand, and at Isca compacted gravel, which as in the *palaestra* was seen as more suitable than flagstones, which would cause hobnailed sandals to slip. The Chester *basilica* had a swimming bath, which was is unusual, although apparently a tradition of *Legio II Adiutrix*, which built it.

HOSPITAL – THE *VALETUDINARIUM*

During the period of the Republic, with armies raised for specific campaigns, the importance of keeping troops fit was not seen to be of high importance. However, during the period of the Principate, the size of the empire grew, yet armed forces, now a career choice, were of relatively small numbers and expensive to maintain, and therefore it was particularly important to keep the troops fit and healthy. Vegetius commented that it was the duty of the officers of the legion, of the tribunes, and even of the commander-in-chief himself, to take care that the sick soldiers were well supplied with proper diet and diligently attended by the physicians. He qualified this with 'however, the best judges of the service have always been of the opinion that daily practice of the military exercises is much more efficacious towards the health of an army than all the art of medicine'. Vegetius insists that permanent or overnight fortifications needed to be in healthy areas without the problems of diseases contracted from marshes or in areas lacking shade. In summer, the military force should start the journey before daybreak to get to their destination before the midday heat. In winter, troops should not be marching through snow and frost by night, and a good supply of firewood should be ensured. It was also recommended that legionaries be exercised inside rather than in the elements, when it cold and wet. It seems that every soldier was given some knowledge of first aid. While the army high command took special care in its choice of recruits, the diet of the troops, opportunities for exercise and its strategies for protecting soldiers from physical harm when on manoeuvres or on campaign, it was the creation of hospitals in the legionary fortresses and auxiliary forts that was a major innovation.

We know little of medical treatments from literary sources. There are medical texts written by Greek and Roman authors, but we have no idea whether or not they were used in training. Aulus Cornelius Celsus, writing his *De Medica* in the first century AD, sums up the works of Greek physicians by identifying the importance of exercise and hygiene, as well as the study of anatomy. However, many of his medicines were extremely toxic, and it is likely that his cures for diseases increased the mortality rate. Though, with the strongest antiseptic available being honey, all of these procedures was dangerous. There certainly was no understanding of the sterilisation of instruments, or the workings of the heart. Most important was the lack of knowledge of infectious diseases, and the need to isolate patients, as the close confines of the barracks were the perfect breeding grounds for viruses. The movements of troops, especially vexillations returning from warmer climates, also increased the risk of the spread of plague and other sicknesses.

Evidence for doctors attached to a legion is rare, although judging by his procedures for removing missiles, javelins and lead sling bullets as well as treating sword damage, Aulus Cornelius Celsus may well have served in action with a legion. There are two doctors identified from altars at Chester, Hermogenes and Antiochus, presumably Greek, but they were possibly the private physicians of the legate. At the auxiliary fort of Vindolanda in northern England, a document suggests that troops could not undertake duties if sick (*aegri*), wounded (*vulnerati*), or had eye problems, but whether these were treated by a doctor or *immunes* is not known. Tombstones and discharge documents from other parts of the empire indicate that there was a hierarchical structure to hospital staff similar to that of the legion itself. Overall responsibility for health provision was part of the duties of *praefectus castorum*, though the day-to-day management of the hospital was the role of the *optio valetudinarii*. There would have been a medical officer (*medici principalis*) of equestrian status, sometimes on a short service commission, who was fully qualified, and a trained physician under whose direction all staff worked. There was also a *medici ordinarii* who may have had centurion rank. All of the other ranks who worked in the hospital, the *capsarii* (dressers) and the *seplasiarius,* supplied the medical ointments and were attached to centuries. While the wound was healing, the *optio convalescentium* ensured the provision of a nourishing diet and the occasional bath. While 'wounded' men would be unlikely in a static legionary fortress away from the frontier (though auxiliaries may have been brought down from forts), there were always going to be accidents while on exercise and in construction of buildings, including the breaking of bones. The baths might have offered massage for the treatment of muscular problems. Some men might have been treated in barracks for some complaints, or treated by their mess mates. Presumably, the tribunes and legate would have had their own arrangements, though they were unlikely to be wounded or hurt through physical activity.

The identification of legionary hospitals is fraught with problems. Hyginus tells us that hospitals did exist in temporary campaign camps, a rectangle of tents best situated between behind the commander's house (*praetorium*), in between the *veterinarium* and the *fabrica* 'so that there may be peace for those convalescing'. However, the placing of a building in a temporary camp says little about the situation and architecture of the *valetudinarium* in a permanent fortress – the building is not always in the same position and this varies greatly between fortresses (although at Inchtuthil the hospital is in the position identified by Hyginus). At Caerleon there have been at least three candidates for the *valetudinarium*, as there are a number of rectangular buildings in the fortress, some in the position given by Hyginus behind the *praetorium*, but, as we have seen, they have also been variously identified as *fabricae* (storerooms). Similarly, at many auxiliary forts what were identified as hospitals have turned out to be workshops.

The 'type-site' for the hospital was identified at Novaesium, present-day Neuss, Germany, where ten (medical) probes were found in one room and four scalpels in others. However, the instruments were all damaged and unsuitable for use. Following this example, all military hospitals are proposed to be similar in plan, with two ranges of rooms arranged around an internal courtyard. The architecture of the hospital has been suggested as

i)   It is entered by a portico that leads to an entrance space.

ii)   A large clearance hall is divided into nave and aisles, with a separate opening to the operating theatre with remains of raised hearths and a small room for sterilising instruments. The operating theatre is presumed to be a room jutting out into the courtyard because of the better quality of the light and fresher air.

iii) Sixty cubicles/wards, one for every century with four men in each on bunks, were arranged on either side of the circulating corridor, and in pairs separated from the next by a side corridor, which was probably a small room for preparations, or to insulate against noise. At Novae, personal possessions, such as *amphora* for olive oil used for lamps or dressing wounds, were found in these spaces. There would have been lavatories (facilitated by chamber pots), a kitchen, a set of baths and two mortuaries, as well as a treatment room containing instruments for surgery.

What often seems to be lacking is a kitchen for preparing food, although this might be evidenced by the raised hearths in the clearance hall, or sustenance might have been

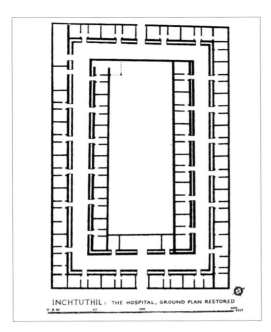

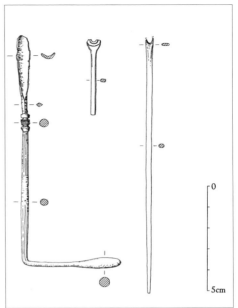

Above left: A reconstruction of the plan of the hospital at Inchtuthil. (*Pitts & St Joseph, p. 196*)

Above right: Surgical instrument's from the Blake Street barracks in York. (*York Archaeological Trust*)

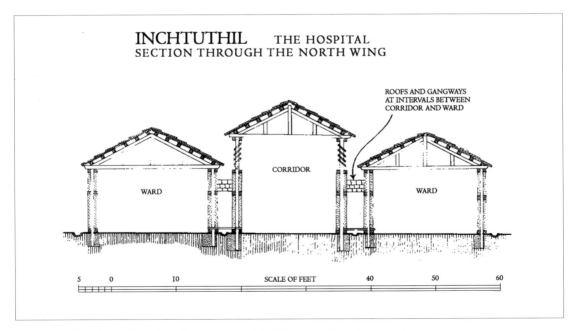

A section through the hospital at Inchtuthil. (*Pitts & St Joseph, p. 197*)

have been brought by the sick soldier's mess mates from his *contubernium*. However, some of the proposed features of these fortress hospitals, such as the provision of wards, sterilisation of instruments needing hearths, and the concept of isolation from infectious diseases, are practices of modern medicine and not all available to the Roman army.

The present accepted location for the hospital at Isca is tucked away between the baths and the *via sagularis*, and protected from the noise and bustle of *via principalis* and *porta principalis sinistra* by a tribune's house. There was no other important structure along the *via sagularis* in this quadrant of the fortress, and this would have ensured minimum traffic. Although there was a row of barracks in front of the hospital, they were '*contubernia* end-on' and therefore limited noise. Being next to the main sewer on the *via sagularis* would have been a disadvantage, although it would have enabled a thorough discharge of human waste from the building. The position of the *valetudinarium* close to the *thermae* and the *basilica* was particularly apposite in many ways for hot waters and massage, however, Seneca's echoing vaulted spaces would have amplified the racket of troops enjoying themselves or exercising hard, especially with the shouting of the officer in command. The bath building with its high bulk would stop cold south-west winds, and it would also shade the courtyard of the hospital from the high noonday sun in summer.

The 'hospital' at Isca was excavated in 1964, with only the area of half of the structure being examined by narrow trenching (i.e. probably less than a quarter of the footprint), and nearly all the walls had been robbed down, or nearly so, to foundation level, and this made identification of the structure difficult. The provision of two water tanks inside the building and two in the courtyard suggested the collection of rainwater was specified as best for invalids. The excavator thought that the small wards outside the corridor were probably grouped in threes in the inner ward, and twos in the outer. Only one room appears to have been heated, and the few traces of drains were unlikely to have carried away foul effluent. No latrines or baths were identified in the courtyard. A similar picture emerges from Inchtuthil, but as there was no reception room, the operating theatre might have doubled up in this case, and it was suggested that one ward per century with four or five beds in a room would accommodate between 5 per cent and 10 per cent of the legion if necessary.

Curiously, it is unusual for medical instruments to be found in the hospitals, and these are often damaged and unsuitable for recycling. This might be explained in the careful curation of important personal tools by the staff, and which would leave the fortress with them. Probes with different shapes at either end, such as a small scoop/blade, *spatulae*, spoons, elevators, tweezers, curved and straight needles clearly have personal toilet as well as surgical or medical usage. Similarly, grinding instruments for herbs that were grown in a hospital courtyard could also be used throughout the fortress. Scalpels, bandage clips, boxes for ointments, glassware, scales, mixing bowls of many sizes, a double-ended blunt fork for dividing muscle tissue, dental forceps, scissors and leg splints have all been identified from modern examples, which itself may be misleading. At Isca, thirty-eight possible medical instruments have been found in the barracks, amphitheatre, rampart areas, the baths and one of the possible workshops. Similarly, no surgical instruments or medicines were found in the hospital at Inchtuthil. However, many personal items of decoration and dress have been found in the narrow rooms between wards at Novae.

While there is considerable controversy about the plan of, and practice in, the fortress hospital, there is some valuable evidence from Novae to support the accepted plan of the structure. In the courtyard, excavators unearthed a podium 2.46 metres by 2.60 metres, with an entrance of small steps which led to two columns in the façade. There was an open passage behind a screen, probably made of wood. Several statue bases were inside the podium, two inscribed to the gods of healing, Hygiaea and Asclepio, and dedicated on behalf of the legion. In front of the façade of this small temple were numerous other, smaller altars and bases for silver statues, clearly the offerings of individual patients expressing their gratitude for being healed. This place was clearly a *sacellum* (a cult area) and points to the sacred nature of the courtyard, at least, if not the whole building. This had echoes of Greek cult medicine, where healing took place in a single location dedicated to a deity.

GRANARIES (*HORREA*)

If either, or both, Corellius Audax and Sentius Paullinus were on granary duty for the day, they may have walked from their barrack block down the *via sagularis*, crossing behind the *porta principalis dextra* to the *horrae*, beyond which were the barracks of other cohorts. The granaries in a fortress were usually close to one of the gates, especially that leading down to the river on which a base was situated and alongside the *via sagularis*. At Caerleon, geophysics has identified the granaries in the modern Priory Field, a location that made access to the rest of the fortress possible without having to enter the high-status administrative areas. *Horrea* also needed to be in a situation where there was enough room for good access and space for vehicles to manoeuvre, preferably alongside the loading platform – there have been examples of cobbling found near loading bays. Presumably, though we cannot be sure, the corn would have been allocated using a measuring vessel of some sort and then put in sacks for transportation to the barracks. Whether it was distributed for the day or week is unknowable, but daily would make storage in the barracks easier and it would not be difficult for the granary workers to deal with the sixty-four centuries. Whichever method was used, what we can be sure of is that the amounts of grain distributed would have been recorded by the clerks to the granaries (*horreorum librarii*), probably in triplicate.

The staple food of the legionary was cereals (as it was for most people from the Neolithic period until the eighteenth century), and it has been estimated that to feed a full fortress for a six-day period would have taken about 8 hectares of wheat, 40 per cent of the size of Isca. The yearly ration of each legionary was around 341 kgs, which would have taken up about 0.45 cubic metres of space, and its storage was of paramount importance to the viability of the legion as an effective fighting unit. A full legion in residence in the fortress theoretically needs 2,016 tonnes a year. Isca had three granaries, as did Deva, although Inchtuthil had six, probably because it was acting as a grain reserve for the auxiliary forts in its isolated command. Sometimes it is suggested that there may have been a policy of keeping a year or so supply of corn in case of emergencies, but with each of the three granaries needing to contain 672 tonnes, this unlikely and there must have been a continuous topping up as they were emptied. There is evidence from the *canabae* of weed seeds from the Mediterranean in a deposit of burnt grain, but this was probably meant for brewing. If these large amounts of grain were not to spoil or be destroyed by accident, the granaries had to be designed, and continually maintained, to combat the most common threats to their contents.

Grain is a particularly difficult commodity to store effectively in bulk for the long term. After harvesting, grain continues to respire, taking in oxygen, giving off heat, carbon dioxide and water. These processes need to be slowed down if the grain is going to survive and be consumed at a later date, and this involves reducing the moisture content and oxygen available, which will ensure that it remains dormant. If the grain gets too hot or wet it will begin to germinate, and dampness will encourage infection by bacteria, resulting in mould and fungi. In turn, insects, such as the saw-toothed grain beetle, the grain weevil and the flour mite would flourish, so the best way to stop these pests was to keep the temperature low. A further menace to bulk-stored wheat supplies are small animals, such as rats and the common mouse, looking for food and shelter. Obviously, birds also would take the opportunity to access this significant supply of food and could become a major nuisance. Each of these hazards were considerable factors in the design of granaries in the fortress.

The fundamental structural requirements of the design must be able to cope with the pressures that grain, acting as a 'semi-fluid', puts upon the walls and the floor of a granary. This means that the foundations for the walls discovered by excavation are substantial and vary from 0.73 metres to 1.30 metres wide. Sometimes, the foundations are so considerable that there is the possibility of a second store, such as a loft for the storage of items such as vegetables, cheese, meat and olive oil. Buttresses, usually in opposite pairs, not only supported the walls but also the roof. To keep the grain dry, the roof needed to be entirely weatherproof, preferably clay tiled, which would also make it fireproof, and any water pouring off the tiles needed to be channelled away from the granaries.

To reduce temperature and moisture content, the granary had to have ventilation for circulating air. This was achieved by supporting the storage area on wooden or stone sleeper walls or columns. Ventilators through the walls, between the buttresses, had grills of iron mesh or timber to keep rodents out, and flagstone floors in the storage space also helped stop rising damp. Keeping the grain dry as it was delivered would have meant providing some sort of *portico* and doors that might form a passageway into the main storage space. Inside the building, storage of the grain in sacks, rather than bins, would help to keep the air circulating, especially if they were placed on wooden gratings of some sort. To discourage insects, especially the dangers posed by eggs, which remain viable for long periods, the walls would need to be made smooth by plastering them and ensuring that the floors were free from holes and cracks. The gables and eaves would have been well plugged to stop birds getting in, and it may be that keeping the building dark would deter birds from flying in.

It is clear that *horrae* probably presented the most complex structural challenges of any building in the fortress, except the baths, even though they have a simple footprint and were not especially impressive from their exteriors.

7

Life and Death Outside the Fortress

Community spaces also existed outside the fortress, which were particularly important in forging loyalty and allegiance to the legion through ceremonies, festivals and entertainment, as well the training of troops to keep them fit.

THE *CAMPUS*

At Isca, the parade ground was thinly metalled and measured 220 metres by 145 metres. It would seem that it was only in existence from AD 140, though considering the importance of the facility, it must have been elsewhere in previous years. Walls have been found on two sides, but as yet the gateways, if they existed, have not been located. Parade grounds give little archaeological evidence because of their size, but also since it was the place where soldiers would have met for recreation, any small pieces of armour would have been picked up and used as scrap, or added to personal kit. We are very much reliant on documentary evidence for our understanding of these spaces.

Josephus, who had fought against the Romans in the Jewish Wars, changed sides and became a Roman citizen, was diplomatically keen to praise the army:

For this people does not wait for the outbreak of war to practise with weapons, nor do they sit idle in peacetime bestirring themselves only in time of need. Rather, they do seem to have been born with weapons in their hands; never do they take a break from their training or wait for emergencies to arise. Their manoeuvres fall no way short, in the amount of energy expended, of real warfare, but everyday each soldier exercises with as much intensity as he would in war. This is the reason why the shock of war affects them so little.

(Jewish War III, 71–107)

However, Vegetius has a more cynical view, seeing the legion as a static, peacetime garrison with well-trained and intelligent men who might get bored and mutiny. He was at pains to convince his patron of the benefits of training for keeping troops busy to stop insurrection.

> They must continually practise manoeuvres, have no leaves, be kept busy with roll calls and parades; they are frequently to be kept occupied for the greater part of the day, until the sweat runs off them, at throwing their weapons, the movements of arms drill, and thrusting and slashing with their imitation swords at the posts; they must also be exercised in running and jumping over ditches; during the summer they must be compelled to swim in the sea or any river that is near their camp; the troops must be made to cut down trees, to march through thickets and broken ground, trim timber, dig ditches, and one party in turn should occupy a site and use their shields to prevent the other party from dislodging them. Thus the troops, legionaries, auxiliaries and cavalry, will be well exercised in their camps.
>
> Vegetius

At Lambaesis, Africa, the parade ground was 2 kilometres from the fortress, measured 200 metres square and was enclosed by a perimeter wall 60 cm thick with two gates. At its centre was a tribunal that was later inscribed with an account of a visit by the Emperor Hadrian in AD 128. From this raised platform he would have watched the training of his troops and gave an *adlocutio* (a speech) for raising morale and developing *esprit de corps* between him and the legion. Hadrian wanted to be seen as not only the commander-in-chief, by inspecting all aspects of the daily administration of the army, conditions in camps, forts, living quarters, weapons and even the records of soldiers, but also to demonstrate his *commilitones* (comradeship) with the *milites*, in some way becoming one of them by marching bareheaded in full armour in the cold or heat, wearing simple military dress, and visiting the wounded, as well as eating simple food and drinking sour wine. He was known to supervise drills, which further emphasised his credentials to criticise and give advice.

The *adlocutio* was an opportunity to convey these messages directly to large numbers of troops in a single speech, and therefore the importance of the *campus* would fit a whole legion. It was an opportunity for the emperor to translate the oath of allegiance into the creation of a true bond with his soldiers. He probably would have delivered his *adlocutio* in full armour, though apparently without his sword, demonstrating that he knew his troops were loyal to their emperor.

He opens his speech (now only partially decipherable) by emphasising the legion's commitment to training, even though they have been involved in moving fortresses:

> For this I would have forgiven you if something had come to a halt in your training. But nothing seems to have halted, nor is there any reason why you should need my forgiving ... you would ...

Hadrian may well have taken part in manoeuvres away from the fortress, as he reports:

> ... work others would have spread out over several days, you took only one day to finish. You have built a lengthy wall, made as if for permanent winter-quarters, in nearly as short a time as if it were built from turf which is cut in even pieces, easily carried and handled, and laid without difficulty, being naturally smooth and flat. You built with big, heavy, uneven stones that no one can carry, lift, or lay without their unevenness becoming evident. You dug a straight ditch through hard and rough gravel and scraped it smooth. Your work approved, you quickly entered camp, took your food and weapons, and followed the horse who had been sent out, hailing them with a great shout as they came back.
>
> Translation by Speidel (2006)

These activities might have been carried out relatively locally, as above Isca was a late Iron Age hill fort, which shows evidence as having been used to practice assault tactics by attacking the remaining empty fortifications. There were 20-mile marches three times a month, and manoeuvres much further away in the Welsh mountains. Llandrindod Common was the base for many practice camps, perhaps undertaken with local auxiliary troops.

OBSERVANCES, FESTIVALS AND CEREMONIES

The calendar of festivals had been established by papyrus documents from military establishments in Egypt, and one of their aims was to demonstrate that the legion was part of a greater society and community. While the legionaries might have worshipped local gods, the gods from their original district, widely known deities such as Mercury, or the new eastern cult of Mithras, the legion was a corporate body dedicated to the Capitoline Triad: Jupiter, Juno and Minerva. There were annual sacrifices to these deities. On 3 January, an ox was sacrificed to Jupiter, and cows to Juno and Minerva. On that occasion, the *votorum nuncupatio* (public statements of vows) was taken by the troops, and a new annual altar was dedicated to Jupiter on the edge of the parade ground with the old one being ritually buried. On 19 March, Minerva was worshipped along with Vesta. The uniformity of the empire was stressed by the anniversary of the foundation of Rome on 21 April, and the reigning emperor's birthday was celebrated, as well as certain deified emperors – Julius Caesar (12 July) and Marcus Aurelius (29 April). Troops were also encouraged to worship abstract gods such as Disciplina and Honos, the Roman god of chivalry, honour and military justice, who was sometimes conflated with the deity Virtus, representing military virtue in fighting bravely.

The *contubernium* was a tight group of legionaries, which made it all the more important to encourage the loyalty of each of these groups to the legion. There had been mutinies that had usually started with the legionaries. Hence occasions were held, which involved assemblies that were purely military and aimed to develop the coherence of the legion. An annual event was the birthday of the Eagle of the Legion

(*Natalis Aquilae*), in the case of *Legio II Augusta*, it was 23 September, the Emperor Augustus' date of birth. On 10 and 31 May, military standards were decorated in a ceremony called *Rosaliae Signorum*. An elaborate ceremony, the *Honesta Missio*, occurred on 7 January to celebrate time-served veterans being honourably discharged, and would have been of great significance to their mess mates. A further, simpler event was the legionaries lining up on the parade ground to receive their money three times a year, forging a link between all the hard work they did and its reward.

THE *LUDUS*

Only two legionary amphitheatres have been identified and excavated in Britain: at Caerleon in the 1920s, and most recently at Chester between 2004 and 2006, although a number of auxiliary forts have smaller examples. The construction of these amphitheatres so soon after the foundation of the fortress suggests that permanent bases had been planned for some time. The earlier fortresses, such as at Glevum/Gloucester or Isca/Exeter show no signs, so far, of having such a facility built in either timber or stone. This was perhaps because both of these campaign fortresses were probably rarely full with legionary troops, and the resources necessary to undertake such a major project could not be spared. The lack of an amphitheatre has connotations of temporariness, as it suggests an underdeveloped sense of community within these sites.

The amphitheatre at Caerleon is still the most completely excavated and displayed example, and typifies the relatively modest legionary amphitheatres of the Empire's frontiers, rather than those of a populous city. It was constructed in AD 80, soon after the foundation of the fortress, which signifies a sense of permanency. This stability was even more evident when the amphitheatre was built outside the safety of the fortress and, as at Isca, it involved the filling in of part of the defensive ditch, resulting in the top tiers of seats overlooking the *via sagularis*. The arena is 41.6 metres by 56.08 metres, and therefore designed as ellipsoid in plan and its floor was of sand, with a box drain underneath. When excavated in 1926–28, it was presumed that a high stone wall around the outside was 9.75 metres above the arena, which would have allowed the seating directly on the banks to be at angle of 25 degrees. However, due to some surgical excavation by George Boon in 1962, it has been shown that the banks held up an open timber framework resting on a stone base, and braced so as to survive even the most enthusiastic foot stamping. Excavations of the *ludus* at Chester suggest that the timber framework was prefabricated elsewhere and then installed in an anti-clockwise direction. Once the superstructure had been completed, the arena was deepened and a wall inserted around it. The available earth was used for burying and stabilising the timber framework, and providing the base for seating. At Isca there were eight entrances, the two main ones opposite each other on the long axis of the ellipse, and the six others giving access to the tiers of seating. The walls of the arena, and possibly the exterior, were covered in whitewash with red lines, simulating stone blocks. Facing each other across the arena on its short axis were two entrances covered with tribunals,

Looking over the main entrance to the amphitheatre at Isca into the arena. At one time it was thought that the external wall was constructed of stone, retaining much higher banks with tiers of seating on them. Now it is thought that the banks (*cavea*) were around their original height and supported a timber framework with rows of seats. The entrance, as with the others, was half covered by a barrel vault, the main supports of which can be seen at the entrance and halfway in. The railed area protects a pivot hole for a revolving internal gate. On the left-hand side the remaining stones of the internal archway have a socket for the sliding gate bolt, as does that on the outer entrance arch supports. (*Author's collection*).

which appear to be used by individuals of high status, such as the legate, tribunes and camp prefect.

It would be interesting to know how status was reflected in the seating. Were the centuries allocated particular areas according to their number and were the centurions all seated together? As we have seen, the legionary troops at the time of the conquest of Britannia were at least 80 per cent from Rome, and even at the end of the century they numbered about 20 per cent, the rest originating in the provinces of Gaul, Spain, the Danube and Balkans. All of these men would have shared the entertainment fashions of the provinces from which they came, and it was part of Roman identity to be enthusiastic and accustomed to *spectacula* (entertainments that involved blood sports) that so fascinated the people of the Empire. This sense of festivity would be complemented by the temporary stalls and booths outside the main structure, selling hot food and souvenir items of gladiators. The design of the amphitheatre at Isca, especially the entrances, demonstrates this complete familiarity of the legionaries with the form of the structure – it was built by and for Roman citizens. The amphitheatre symbolises how the community and imperial institution of the army was bound together by a military ethos, which could ultimately be traced back to Rome.

Above: One of the opposing entrances on the short axis of the arena, which may have had boxes for privileged users above. The stairs behind would have been for access to the banks and seating. (*Author's collection*)

Left: Members of the audience would have turned right at the bottom of the stairs, and climbed onto the banks through the arch and entered. Access to the arena would have been through a gate in the forefront.
(*Author's collection*)

These military communities, existing in an alien environment, were standard bearers of the imperial mission, and the legionary amphitheatre was a symbol of Roman power.

Of all the Roman characteristics adopted by the peoples of Britannia, there was a particular lack of enthusiasm for the amphitheatre. Those that did exist are rudimentary, with a basic bank erected, and just a retaining wall around the arena and no back wall to give support to high terraces, which may have required engineering resources only available in the army. However, civic buildings in a town such as the *forum* and *basilica* demonstrate the ability to create large, architecturally refined structures, and perhaps the whole concept of these types of spectator sports may have been completely foreign to the indigenous inhabitants of Britannia.

What happened in the legionary amphitheatre? There were clearly advantages in being able to accommodate the whole legion in a compact space where the individual was close enough to hear speeches, experience ceremonies or weapon demonstrations. This made the *ludus* an important military resource. However, these types of activities, by their very nature, leave few archaeological traces, and the evidence we have for the use of the amphitheatre points directly to a place of entertainment, especially the *spectacula*.

There were two main aspects of *spectacula*: the *venationes* and the *munera*. The first might have involved hunts of beasts by men, beast by beast, or men by beast (the *damantio ad bestias* where a condemned man was executed by being shown to be like a beast). At Deva, in the centre of the arena, there was a large stone block with a lead plug on the top surface, which secured into position an iron staple or ring acting as a tethering block. This might have been to keep the animals in a central position, otherwise they may have instinctively run for the arena edges. At Isca, the arena walls were capped by limestone blocks curved to an overhang and topped by a metal rail to stop wild beast climbing into the spectators (as happened in August 2010 at Tafalla in Spain, where a bull climbed into the tiers of seats and badly injured three people). The organisation of the main entrances, with two parallel wooden doors on pivots that could be opened one at a time if need be, suggests the management of animals before they entered the arena. There is a record of a century of *Legio I Minerva*, based in Bonn, catching fifty bears for the arena in a six-month period.

It has been thought that Isca was too far away from the centres of population for famous trained gladiators to travel. However, there is evidence at Isca for *munera* (gladiatorial displays), as underneath one of the tribunals is a space with stone benches, suggesting a waiting place for contestants.

There are also two decorated stones found in the amphitheatre, indicating the frequent presence of the *retiarius* (who fought with a trident and net) and his traditional enemy the *secutor* (a swordsman with a shield). On one stone there is the image of a trident and the shoulder guards (*galerus*), worn by a *retiarius*, and palm fronds that indicate victory. The other has a central rosette flanked by another *galerus* and a ladder-like object, which may represent the laminated, padded metal guard worn on the sword arm of the *secutor*.

Also found at Caerleon is evidence of the deities normally associated with gladiatorial games – Nemesis, the distributor of good or bad fortune, of success or failure, of life and death (often equated with Fortuna), and Mercury who would conduct the souls of the dead to the Underworld. There is a possible shrine to Nemesis at the western main entrance, and a *defixiona* (a curse tablet of lead) has been found, which begs for divine retribution: 'Lady Nemesis, I give thee a cloak and a pair of boots. Let him who wore them not redeem them (unless) with his own blood.' It is a typical curse where Nemesis was given power over a competitor through an individual's possessions, with the only way of retrieving them being through his injury or death.

There was more to these gladiatorial combats than just the entertainment of watching someone die a bloody death, as the displays also provided an example of courage that would inspire the legionaries. The soldiers would have seen military virtue (*virtus*) in its purest form, as the gladiator showed exemplary courage, skill at arms, unconcern for wounds and the contempt for death that gives the ability to fight without complaining. Another subtle message was contained in the gladiatorial equipment that imitated barbarian styles, and alerted the legionary to the fact that he might have to fight someone who used different strategies from those for which he was trained.

In many ways, the amphitheatre encapsulates the many themes that characterised life in a Roman legionary fortress: the importance of the links with Imperial Rome, the fighting spirit and skills of the men, and the gathering of a community in a public place to forge identity and enhance leisure. Had there not been loyalty to Rome, and a settled, well-housed and well-fed community, it would have been very dangerous to have men with such fighting spirit and skills in the same place. Vegitius warns of the links between too much leisure and mutiny. When Hadrian visited Britannia in AD 121/22, he would have been wise to have seen his legions, especially as he was going to charge them with going north to build a wall that was to define the limits of a static Empire, and that would take the *Legio II Augusta* away from the Isca for a couple of decades. Whether such an *adlocutio* would have taken place on the parade ground, or in the amphitheatre, cannot be known, but it would have to emphasise loyalty to the Eagle, as there would be the breaking of many personal bonds that the men had formed locally. Life in the fortress would have been very different, and with the remaining centuries or cohort caretaking the base, the amphitheatre would be low on their list of structures to be maintained.

MILITARY STRUCTURES OUTSIDE THE FORTRESS

The area outside the fortress was the site of other military activities, besides the *campus* and *ludus*. Some structures must have been devoted to supplies and communications, such as quays and warehouses, and others for functions that could not be fitted into the tightly packed fortress. There may have been annexes for military animals or storage wagons, or a temporary camp for auxiliary troops visiting the base. High-standing official visitors, such as the governor of the province, would have been housed in the legate's quarters. Official travellers and imperial messengers would have been accommodated in a *mansio*, a superior structure with rooms around a courtyard, and the provision of a bath building for which there are several candidates in the settlement at Caerleon. Except for the possible unofficial barrack room shrines, the official

Above: The opposite entrance on the short axis. There is a possible shrine to Nemesis on the back wall, and the stone seat suggests a waiting area for combatants. (*Author's collection*)

Below: The amphitheatre at Chester under excavation 1982–86. The two rear walls of the bans can be seen, the outside one being the later. During the first excavations in 1960, the square structure in the inner wall was thought to be a shrine to Nemesis. The slots held the timbers of the seating supports. (*Copyright of Cheshire West and Chester Council and English Heritage*)

religious beliefs of the empire were centred on the *principia,* and the worship of other deities had to be accommodated outside the fortress. There are inscriptions on stone to Jupiter Dolchensis and Mithras, both favoured cults among soldiers, and also Diana and the Matres (the Mothers). The location of the temples dedicated to these deities is, as yet, unknown.

The most startling discovery of the last decade was in 2008, when students from Cardiff University were undertaking surveying activities and discovered a large, important building between the amphitheatre and the river, a location that would have been the first sight of the fortress as a ship came around the large meander of the river, which is now eroding the remains. The lines of the walls were confirmed using geophysics, with many of the structures being just a few centimetres below the turf, often appearing as undulations of the ground surface, which had previously been interpreted as the spoil from the excavation of the amphitheatre. A large courtyard structure with a monumental appearance, and a rectangular feature in its centre was investigated along with other adjacent structures. Several functions have been proposed: warehouses, a training hall for troops with holes in the ground for holding stakes to practice swordsmanship, a market where supplies for the legion were landed from the *prata* or other parts of the Roman world, an area for corralling cattle or sheep and bringing meat, grain, wool, leather and timber into the fortress. Two further functions have been suggested, which relate to the activities that we presume took place outside the fortress: a school for gladiators, and the rather more refined buildings with heating and baths acting as *mansiones* for visitors of high status.

THE *CANABAE LEGIONES*

A centurion newly transferred to the *Legio II Augusta* who was sitting high up in the amphitheatre would have immediately made sense of the organised plan of the fortress. It would have been very similar to the one that he had left, but turning around and looking over the land between the military base and the river he would have seen something very different – the sprawl of the *canabae* and its inhabitants (*canabenes*). Every permanent fortress developed a neighbouring civilian community, often as soon as it was founded. Finding its extent, or trying to discover its history, is problematic as it would have developed incrementally, initially without a plan and on more than one side of the fortress. Not having the predictable plan of the legionary base within its four walls, it is difficult to ascertain the limits of a *canabae*, and again the presence of medieval buildings at Chester and York hide the remains of the settlement. The layout of the *canabae* was probably rather irregular until the construction of the parade ground in around AD 140, and then the settlement's plan seems to have been set out parallel to this new axis, possibly under the orders of the legate. The *canabae* could be a large and thriving settlement of several thousand civilians, the relationship with the fortress being mutually beneficial. It stood on garrison land and the legate had control over the area. In turn, the settlement increasingly supplied the legion with recruits, but also more ephemeral services needed by young men who were living in a totally male environment and wanting to spend their pay. There were sellers of luxuries and entertainments, taverns and women. The wives and families of serving soldiers, who officially were contracted not to marry but entered common-law relationships, would have lived in the *canabae*, as well as veterans and their families who wanted to stay close to the routines of twenty-five years, and the servants and slaves of both. This relationship between fortress and the occupants of the *canabae* can be seen in evidence for non-military persons being admitted to the amphitheatre, and the traces of women and children in the fortress baths. There is evidence of a variety of tradesmen and craftsmen, some probably just making an opportunistic living while others were contracted to the army. There is no indication of whether there was a local population, or if some came from further afield following the legion with some perceived sense of loyalty. When vexillations of the legion moved away there may well have been enough of a civilian population to maintain the *canabae*, but we do not have the evidence of this as we are lacking a history of the settlement. One thing that must be considered is that both the military and the civilian would have been a distinct community, sharing a broadly similar attitude towards the local people, and would have been conspicuous as intruders, especially in the early years of the occupation.

It would seem from evidence of *canabae* along the northern frontier of the Empire, especially from fortresses along the Rhine and Danube, where there has been more research, that these settlements were carefully zoned, with domestic housing separated from industry – pottery, tile and glass manufacture –and probable markets. Because of the lack of excavation at Caerleon, and therefore information, it is difficult to identify these zones. There are a range of buildings, from the sophisticated courtyard houses with their baths, hypocausts for central heating, painted plaster and mosaics, to strip houses with the doorway on the narrow side facing the street that they lined. It certainly appears that the area on the amphitheatre side outside the *porta principalis dextra* was of higher status than on the opposite side of the fortress. This lower status part of the *canabae* outside the *porta principalis sinistra* seems to have had many strip houses, as well as evidence for rubbish disposal, marshy wetlands, and were inhabited by people of the edges of society. There appears to have been subsistence farming carried out between the fortress walls and the line of the houses. As around many fortresses, there is no significant settlement outside the *porta decumana*, at the area beyond the back of the base.

How much the *canabae* was managed by its inhabitants is difficult to know, but there are examples elsewhere of some gaining chartered status (*municipia*), being self-governed and with a *basilica* for the *ordo* (council leaders) and a forum for selling. Others, such as Eboracum at York, developed into *colonia* (a settlement for retired legionaries), with their own administrative arrangements and high status. At Isca this may not have been the case, as the *civitas* capital of the Silures at Venta, now the village of Caerwent about 8 km from Isca, was established, possibly with the encouragement of the legion and help from its surveyors, and would have been a potential limitation of the growth of the fortress *canabae*.

THE *PRATA*

The *canabae* was on land that was part of the *prata,* or after the reign of Hadrian more commonly referred to as the *territorium*. It was owned by the legion and under the direct control of the legate. The earlier name of *prata* (pasture) demonstrates the importance of providing the grazing for the legion's animals rather than referring to cultivated land, which was also an important function of the area. With a mule for every *contubernium,* and therefore sixty per cohort, this adds up to 640 per legion. Oxen would have been used for baggage and other heavy duties, as well as the 120 horses for the legion's own cavalry (*equites*), and it is likely that the mounts of the higher officers were kept inside the fortress for ease of access. Since the legion aimed to be as self-sufficient as possible, there were probably several thousand cattle in herds managed and brought to camp by legionaries (*pecurii*), who were also responsible for the quality of the meat. It is likely that much of the agricultural land was farmed by veterans or civilians, as tenant farmers, producing directly for the legion or handing over surplus supplies. Near Isca, on the opposite bank of the river at Bulmore, a small village has been found, and a similar settlement at Heronbridge near Deva may represent the homes of veterans.

There would be variations in soil, geology and vegetation in the *territorium*, and so the legion may not have all its land in one block but instead spread over a wide area, perhaps in discrete sectors that would ensure natural resources of stone, clay and timber were readily available and easily accessible. We are aware of a planned system of reclamation of marshy land along the River Severn, with sea banks being constructed to control tides and run-off from the hills behind. There are also indications that the legion was involved in the mining of lead ore and coal. Perhaps because of the amount of flat land alongside the rivers of the coastal plain and their continual flooding, much of the agriculture and industry associated with Isca had been deeply buried by alluvium from the bordering rivers. We don't know the location of a stores compound, like the one recently discovered in Exeter, or the legionary tile and brickworks, such as that at Holt, Deva, where a barracks compound, bathhouse, possible officer's residence and eight kilns were constructed during the period of the fortress' transformation from timber to stone buildings. Identifying the limits and area of the *territorium* is not possible on present evidence, but David Mason has produced some useful models that suggest that the land belonging to the successive legions at Deva might have been as much as between 200 and 350 square km, and that of *Legio II Augusta* up to 375 square km. The area may still have been used in the legion's absence from its fortress, as many of the tasks that took it away were undertaken in regions of poor agricultural land, such as along Hadrian's Wall, and supplies of food would still be needed. Whatever the area, it is certain that the foundation, repair and everyday life of the fortress would have made a huge impact on the surrounding landscape, some of the consequences we may still be living with today.

A settlement around an auxiliary fort (*vicus*) would have similar functions to the *canabae,* but on a much smaller scale, perhaps with the involvement of a greater proportion of local people. Since the majority of these forts were in unsuitable territory, as far as agriculture was concerned, they no doubt relied on supplies from their parent fortress, and any areas of land around the forts would also be under the command of the legate of the legion whose auxiliaries they were. There is evidence in terms of earthworks still in the landscape that both types of troops came together to perform exercises that would keep them up to the expected standard, and fighting in tandem against the enemy.

DEATH IN A ROMAN LEGIONARY FORTRESS

In the end, those who lived in a legionary fortress would have died, though many in the legion would have been buried away from base on a detachment elsewhere, or would have left after their twenty-five-year service. Burial was always outside of the fortress for health and spiritual pollution reasons, and it would seem that the main cemetery at Isca was alongside the road leading from the *via decumana* at the rear of the fortress. This had nothing to do with the low status of the area, but everything to do with the cemetery, 1 km in length, being on a hillside that dominated the whole fortress and the valley around it. The cemetery appears to be planned, but whether for soldiers or civilians it is difficult to ascertain. Of the 121 cremations excavated at the Abbeyfield site in 1992, where it was possible to allocate gender, the remains were predominately of males (a possible fourty-four individuals), but there was little evidence connected with military in the form of headstones or grave goods, even though twenty-seven graves contained hobnails. These were all simple burials, but as legionaries belonged to burial clubs it would be expected that their mess mates would have ensured a gravestone for the deceased soldier, as is often found elsewhere. Females (a possible eight) and two children were also present, though this may not be the pattern throughout the unexplored parts of the cemetery. The age range also did not indicate the main cemetery for troops, with two children under the age of puberty, two individuals between fifteen and eighteen years, three between eighteen and twenty-five, and five between twenty-five and thirty years. Eleven individuals were thirty years and over, one in their forties and one who was at least fifty.

The burials seem to have been predominately from the first or early second centuries, and the cemetery was in use until at least the early third century. During the earlier period it was cremation, rather than inhumation, that was the major way of dealing with the body of the deceased, and we have enough evidence to reconstruct the proceedings. Most information of the earlier stage of death ceremony comes from literary sources. There appears to have been a common ritual for all, although the richer you were the more elaborate the ceremonies and memorials. When death was thought to be near, the family gathered and the closest relative gave the last kiss to catch the person's soul. Then the family shouted grief and called the name of the deceased. The body was washed, anointed, dressed and laid out, and carried to the cemetery, the relatives dressed in black. There may have been pipers, trumpeters and horn blowers playing (would the legion have provided music for its veterans?).

It is at the cemetery that archaeological evidence can aid our knowledge at what happened at particular localities. The body was burned at the place for cremation (*ustrinum*), and gifts and personal possessions would have been put on the pyre. While no evidence of the pyre was found at Caerleon, the presence of many partly burnt bones at Trentholme Drive, York, from different individuals, and funerary offerings, indicates a pyre. The pyre was put out with wine, and the remains collected and put in some sort of container – in Caerleon cists, urns made of pottery, either jars or beakers, have been found. There was some evidence of caskets and boxes, and where absent, bags made of leather or cloth must have been used. Generally objects were placed as grave goods for the afterlife, but at Caerleon few grave goods were found. These include two unburnt urns, which suggest supplies of food, a spindle whorl in a woman's grave, a spearhead accompanied a child (who might have wanted to be a legionary in later life?), and lamps to light the dark journey after death, which were found inverted into the grave. There were no toys put with children, as have been found elsewhere. After the burial, a feast was held at the graveside (*silicernium*) and bones from pigs, domestic fowl, and smaller birds, were found with eleven burials and might represent this final part of the ritual. Finally, a tombstone might be raised, though if it was a marker of wood it would not have survived.

> To the memory of TITUS FLAVIUS NATALIS, a veteran. He lived 65 years. Set up by FLAVIUS INGENUINUS and FLAVIUS FLAVINUS his sons and FLAVIA VELDICCA his wife.

Veldicca is a Celtic name, probably a girl from the *canabae*, but his sons would have been Roman citizens.

> To the memory of QUINTUS JULIUS SEVERUS, from DINIA veteran of the LEGIO II AUGUSTA. His wife set this up.

Dinia is the modern Digne town in Gallia Narbonensis, Gaul, the recruiting ground of *Legio II Augusta*.

> To the memory of TADIA VALLAUNIUS who lived 65 years and of TADIUS EXUPERTUS her son, who lived 37 years and died [served?] in the German campaign. Her daughter TADIA EXUPERATA set this up in loving devotion to her mother and brother near the tomb of her father.

Tadia Exuperata is also a Celtic name, and the tombstone indicates that part of the legion was away in another province of the Empire.

In many ways, the discovery of burials is the closest we can get to the individuals who lived in and around Isca. These touching messages, evidence of Celtic women, the recruiting of legionaries and their service abroad, supports much of what has been said in this book – it is the loving devotion of family members that is often invisible in the archaeological and literary sources, and these tombstones add that important, emotional dimension to life in a Roman legionary fortress.

Bibliography

THE ROMAN ARMY

Bishop M. C., and J. C. Coulson, *Roman Military Equipment from the Punic Wars to the Fall of Rome* (Oxford: Oxbow, 2005).

Campbell, B., *The Roman Army 31BC-AD337: A Source Book,* (London: Routledge, 1994).

Davies, R., P. D. Breeze, and V. A. Maxwell, *Service in the Roman Army* (Edinburgh: Edinburgh University of Press, 1989).

Erdkamp, P. (ed.), *A Companion to the Roman Army* (Oxford: Blackwell, 2010).

Gilliver, K., *The Roman Art of War* (Stroud: The History Press, 2001).

Goldsworthy, A., *The Complete Roman Army* (London: Thames and Hudson, 2011).

Mattingly, D., *An Imperial Possession: Britain in the Roman Empire* (London: Penguin/ Alan Lane, 2006).

McNab, C. (ed.), *The Roman Army: The Greatest War Machine in the Ancient World* (Oxford: Osprey Publishing, 2013).

Shirley, E., *Building a Roman Fortress* (Stroud: The History Press, 2001).

Wilmott, T., *The Roman Amphitheatre in Britain* (Stroud: Tempus, 2008).

LEGIONARY FORTRESSES IN BRITAIN AND BEYOND

Bidwell, P. T., *Roman Exeter: Fortress and Town* (Exeter: Exeter Museums, 1980).

Bishop, M. C., *Handbook to Roman Legionary Fortresses* (Barsley: Pen and Sword, 2012).

Burnham, B. C., and J. L. Davies, *Roman Frontiers in Wales and the Marches* (Aberystwyth: RCAHMW, 2010).

Gascoyne, A., D. Radford, and P. Wise, *Colchester, Fortress of the War God: An Archaeological Assessment* (Oxford; Oxbow, 2012).

Knight, J. K., *Caerleon Roman Fortress, Third Edition* (Cardiff: CADW, 2003).

Mason, D. J. P., *Roman Chester: Fortress at the Edge of the World* (Stroud: The History Press, 2012).

MV InstytutArcheologii Novae: Research in the Headquarters. URL: http://www.provinces.uw.edu.pl/novae_en.html (Accessed 12 July 14).

Ottoway, P., *Roman York: Revealing the Past* (Stroud: The History Press, 2004).

EXCAVATIONS AND FIELDWORK REFERRED TO IN THE TEXT

Caerleon Research Committee. URL: http://www.cf.ac.uk/hisar/archaeology/crc/our-history.html (Accessed 12 July 14).

Chester Archaeology. URL: http://www.cheshirearchaeology.org.uk/?page_id=193 (Accessed 12 July 14).

Davey, P. J., *Chester Northgate Brewery Phase One Interim Report* (Chester: Grosvenor Museum Excavations, 1973).

Ward, S., and T. J. Strickland, *Excavations on the site of the Northgate Brewery Chester 1974/75. A Roman Centurion's Quarters and Barrack* (Chester: Chester City Council Grosvenor Museum, 1978).

York Archaeological Trust. URL: http://www.jorvikshop.com/category.php?cat=16&page_number=1 (Accessed 12 July 14).

The Roman legions were the formidable, highly organised and well-disciplined backbone of the Roman army, vital to maintaining order and control of the borders of the Empire and its subjugated peoples. The fortresses that were the bases of the legions reflected their values: purposeful, hierarchical and an intimidating display of Roman culture.

But what was it like to live in one of these fortresses? What was the everyday experience of the legionary, centurion and commander? *Life in a Roman Legionary Fortress* provides a fascinating insight into the inner mechanisms of the *castrum* and the people who maintained it. Using the fortresses at Chester, York, Caerleon and across the Empire, Tim Copeland reconstructs the complex workings of these legionary camps and provides readers with the archaeological and literary evidence that gives us an insight into life behind the high walls.

AMBERLEY   £14.99

ISBN 978-1-4456-4358-8

9 781445 643588

www.amberley-books.com